W9-AVS-807

MANUFACTURING

GLOUCESTER COUNTY LIBRARY
389 WOLFERT STATION ROAD
MULLICA HILL, NJ 08062

Mc Graw Hill **Education**

Bothell, WA • Chicago, IL • Columbus, OH • New York, NY

Image Credits: Cover Photo: PhotoLink/Getty Images.

www.mheonline.com

 **Education**

Copyright © 2012 by The McGraw-Hill Companies, Inc.

All rights reserved. Except as permitted under the United States
Copyright Act, no part of this publication may be reproduced or
distributed in any form or by any means, or stored in a database
or retrieval system, without the prior written permission from the
publisher, unless otherwise indicated.

Send all inquiries to:
McGraw-Hill Education
130 East Randolph Street, Suite 400
Chicago, IL 60601

ISBN: 978-0-07-661076-1
MHID: 0-07-661076-4

Printed in the United States of America.

3 4 5 6 7 8 9 RHR 17 16 15 14 13 12

The *McGraw-Hill* Companies

CAREER COMPANION
MANUFACTURING

CONTENTS

Copyright © The McGraw-Hill Companies, Inc

TO THE STUDENT

EXPLORING AND PREPARING FOR A CAREER IN MANUFACTURING

This resource booklet is designed to introduce you to the manufacturing industry. It will tell you about the variety of jobs in the industry and how to build a career in this field. It will also provide the opportunity to practice the skills that will help you succeed in the industry. Explore the manufacturing industry and practice the skills presented to help you decide if this industry is right for you.

Finding a job that interests you is the first step in managing your career. To be successful, however, you'll need to explore many job and career possibilities. What if your goals change? What if there is a shift in the labor market or the economy? You may need, or want, to change jobs or even careers. By developing your transferable skills, such as speaking, writing, organizing, planning, and problem-solving skills, you will make yourself a more valuable employee and be able to cope with changes in the labor market. The more transferable skills you develop, the greater your chance of success at any job.

When considering a career in the manufacturing industry, it is important to understand the realities of the industry. Which jobs have the strongest growth? Which offer good opportunities for advancement? Which jobs align most closely with your own abilities and interests? Are there many jobs available in your area?

Keep these questions in mind as you read Part I of this Career Companion booklet. When you have finished, refer back to them and see how many you can answer. Do the answers make you more or less likely to want to work in this industry? If you feel this industry may be right for you, work your way through the practice questions in Part II. Using real world situations, they will help you begin preparing for any career in the manufacturing industry.

Copyright © The McGraw-Hill Companies, Inc

EXPLORE

This section of *Career Companion: Manufacturing* will introduce you to the manufacturing industry.

You will explore the following topics:

THE MANUFACTURING INDUSTRY

MANUFACTURING JOBS

BUILDING A CAREER IN MANUFACTURING

EDUCATION AND TRAINING

WORKING IN THE MANUFACTURING INDUSTRY

INDUSTRY TRENDS

CAREER RESOURCES

After exploring this industry, you will be able to answer the following questions:

- What kinds of jobs are available in manufacturing?
- How can I match my skills and interests with the right job?
- What are the training and education requirements for the job I'm interested in?
- What are some important skills needed to work in manufacturing?
- What is the work environment like?
- What factors affect trends in the industry?

As you read this book, think about whether the careers described are right for you.

THE MANUFACTURING INDUSTRY

Manufacturing is the process of making products by hand or by machine. Most products you use every day come from the manufacturing industry. For example, apparel manufacturers produce the clothes you wear. Breakfast cereal is created at food processing plants. Manufactured products can be as simple as paper clips or as complicated as robots and computers. Manufacturers create their products by hand or by using highly specialized machinery, which they purchase from other manufacturers.

The United States is the world's largest manufacturing economy. It produces products worth $1.6 trillion each year. Manufacturing includes 18.6 million jobs. This is about one in every six jobs in the private sector, or private and non-government companies.

Career Pathways in the Manufacturing Career Cluster

A **career cluster** is a grouping of jobs and industries based on common characteristics. A **career pathway** is an area of focus within a career cluster. Each pathway contains a group of careers requiring similar skills as well as similar certifications or education. The manufacturing career cluster is divided into six main career pathways:

- **Production**
- **Manufacturing Production Process Development**
- **Maintenance, Installation, and Repair**
- **Quality Assurance**
- **Logistics and Inventory Control**
- **Health, Safety, and Environmental Assurance**

Companies in the manufacturing industry fall into three broad categories. Producers make and sell a product directly to the public or to retail merchants. Suppliers are manufacturing businesses that make parts or components. They then sell these parts to other manufacturers. Distributors sell the products of suppliers.

Copyright © The McGraw-Hill Companies, Inc.

Copyright © The McGraw-Hill Companies, Inc. PHOTO: Noel Hendrickson/Photodisc/Getty Images.

There are ten general areas of manufacturing, each making a different type of product. These areas include aerospace, chemicals, computer and electronic products, food, machinery, motor vehicles and parts, pharmaceuticals, printing, steel manufacturing, and textiles and apparel. Job opportunities are available in each of these areas.

PRODUCTION

Workers in the production pathway include people who work on the shop floor of manufacturing plants. These workers may use machines to make electronic parts. They may construct or assemble housing. They may weld metal, or they may print books.

Machinists use tools like lathes and grinders to produce precision metal parts. Foundry workers melt and mix metals and pour them into molds. Sheet metal workers shape and weld flat sheets of metal into large products, such as heating ducts and drainpipes. Tool and die makers create metal parts used to form all kinds of products, from wood furniture to ceramic bowls.

This pathway is not limited to making products. Hydroelectric plant technicians make sure their plants are running smoothly. Other workers operate chemical and nuclear plants. Millwrights install, move, and take apart heavy machinery. Boilermakers assemble, maintain, and repair boilers and other vessels that hold liquids and gases.

Workers in this pathway must be good at working with their hands. They often must be physically strong. They should have good problem-solving skills and be able to work well on their own. Many workers in this field take part in apprenticeships or train at community colleges.

MANUFACTURING PRODUCTION PROCESS DEVELOPMENT

Employees in this career pathway supervise product design and the design of the manufacturing process itself. They make sure that products meet or exceed customer expectations. They also monitor the materials used to manufacture products.

A variety of engineering jobs are included in this pathway. These include design engineers, electronics engineers, and industrial engineers. Engineers design new products, using computers to analyze and test their designs.

Production	Team Assemblers	250,900
	Solderers and Brazers	126,900
	Welders, Cutters, and Welder Fitters	126,300
	Purchasing Agents	118,600
	First-Line Supervisors/Managers of Production and Operating Workers	91,900
	Machinists	55,600
	Computer Controlled Machine Tool Operators, Metal and Plastic	36,900
	Extruding, Forming, Pressing, and Compacting Machine Setters, Operators, and Tenders	29,600
	Manufacturing Production Technicians	18,500
	Aircraft Structure, Surfaces, Rigging and System Assemblers	13,400
Manufacturing Production Process Development	Industrial Engineers	85,400
	Industrial Production Managers	54,700
	Power Plant Operators	12,200
Maintenance, Installation & Repair	Maintenance and Repair Workers, General	357,500
	Production, Planning, and Expediting Clerks	74,100
	Industrial Machinery Mechanics	62,400
	First-Line Supervisors, Managers of Mechanics, Installers and Repairers	36,500
	Telecommunications Equipment Installers and Repairers	35,600
	Security and Fire Alarm System Installers	27,800
	Medical Equipment Repairers	23,200
Quality Assurance	Inspectors, Testers, Sorters, Samplers, and Weighers	77,900
	Quality Control Systems Manager	54,700
Logistics and Inventory Control	Logistics Analysts	41,900
	Cargo and Freight Agents	40,300
Health, Safety, and Environmental Assurance	Environmental Engineers	27,900
	Health and Safety Engineers	9,200

Source: *O*NET Occupational Network Database*

Copyright © The McGraw-Hill Companies, Inc.

There is a variety of other jobs in this pathway. Process improvement technicians work to find the most efficient processes for creating a product. Production managers coordinate all the people and equipment involved in the production process. Precision inspectors, testers, and graders make sure the products meet required quality standards.

Workers in this field must have good problem-solving skills and mathematical ability. They must exercise good judgment and have a detailed knowledge of the production process. Many jobs in this field require a bachelor's degree.

MAINTENANCE, INSTALLATION, AND REPAIR

This career pathway is wide-ranging. It includes all jobs involved with installing, maintaining, and repairing a broad range of devices. These devices can be small, such as cellular phones, computers, and home security systems. Or they can be large, such as nuclear power generators and satellites. This field is vital to every area of the manufacturing industry.

Jobs in installation and maintenance require highly developed technical skills. These skills may be learned through apprenticeships, technical schools, community colleges, and on-the-job training. Workers in this career pathway need to understand electrical wiring and components. They must also be able to diagnose and repair complex systems.

QUALITY ASSURANCE

Employees in the quality assurance pathway make sure that standards are met and procedures are followed. They are responsible for fulfilling the performance requirements that customers expect from various products and services. Some employees in this pathway monitor and maintain the quality of parts. Others inspect raw materials to see that they meet specifications. Still other employees measure and test products and parts to make sure they will satisfy the customer.

Jobs in this career pathway include food science technicians, medical and health services managers, aerospace engineers, and managers of computer and information systems. Employees may have titles such as inspector, tester, sorter, sampler, and weigher.

Copyright © The McGraw-Hill Companies, Inc. PHOTO: Tetra Images/Getty Images.

Good communication skills are required in this pathway. A quality control systems manager, for instance, meets with the marketing and sales departments of a company to define client expectations. He or she supervises the tracking of test results and product defects. A manager in this field oversees supervisors, inspectors, and laboratory workers engaged in testing activities.

LOGISTICS AND INVENTORY CONTROL

Employees in this pathway move products and materials around. Jobs in the field include storage and distribution managers, shipping and receiving clerks, cargo and freight agents, and transportation managers. Employees move raw materials to the production line, unload trucks with raw materials, and communicate with traffic managers. Organizational skills are very important in this pathway.

A logistics analyst, for example, works with data that tells whether a product is available, how well the production process is working, and how the product is being transported to the next location. He or she meets and communicates with management teams to make sure the materials required for production are available and are not too expensive. People involved in logistics make sure deliveries are made on time and that all orders are fulfilled. They keep track of inventory, the amount of the product available to be shipped. They also manage systems that make sure that the price of a product takes into account the cost of producing and shipping it.

HEALTH, SAFETY, AND ENVIRONMENTAL ASSURANCE

Employees in this career pathway ensure the safe use of workplace equipment. They design and install safety procedures for new production processes. They carry out health and safety investigations. Many jobs in this pathway focus on preventing health problems. People in this pathway train other workers in health, safety, and environmental issues.

Environmental science and protection technicians often perform laboratory and field tests to monitor the impact of manufacturing firms on the environment. They investigate sources of pollution and other dangers to peoples' health. Skills needed for this job include a knowledge of biology, chemistry, computers, and public safety and security.

Copyright © The McGraw-Hill Companies, Inc. PHOTO: Monty Rakusen/Cultura/Getty Images.

Manufacturing Industry Outlook

Industry outlook refers to the projected job growth or decline in a particular industry. According to the Bureau of Labor Statistics, growth will be slow in the manufacturing industry in the years up to 2018. The greatest growth will be in the areas of high-tech equipment and pharmaceuticals. Other areas will see declines.

MANUFACTURING JOB OUTLOOK BY CAREER PATHWAY 2008–2018

PATHWAY	JOB OUTLOOK
Production	• Automation of processes will cause some occupations to grow slowly or decline. • Jobs with skills that cannot be automated will remain strong. • Welders, tool and die makers, machine operators will experience growth.
Manufacturing Production Process Development	• Industrial engineering jobs are expected to grow by 14 percent. • Production managers will see slow growth due to automation. • Labor relations jobs will be highly competitive due to a large number of applicants. • Opportunities for power plant experts should grow for qualified applicants as many workers retire.
Maintenance, Installation & Repair	• Job growth and high turnover should result in good job opportunities for maintenance and repair workers. • Job openings for millwrights and machinery workers are expected to grow slowly, but applicants with skills in machine repair should have strong prospects. • Employment for medical equipment repairers should grow faster than average.
Quality Assurance	• Employment of quality-control inspectors will decline slightly.
Logistics & Inventory Control	• There will be a stronger than average demand for workers in logistics. • Economic growth and popularity of same-day and next-day shipments will increase demand for cargo and freight agents. • Demand for industrial truck and tractor operators will decrease at large establishments, which will use more automated systems.
Health, Safety & Environmental Assurance	• Demand for environmental engineers is expected for grow very quickly, rising 31 percent. They will be needed to help companies comply with governmental regulations and clean up environmental hazards. • Health and safety engineers should see average demand. They will be most needed to ensure that new production and processing technologies are safe for workers.

Source: US Department of Labor, *Career Guide to Industries 2010–2011* and *O*NET Occupational Network Database*

Copyright © The McGraw-Hill Companies, Inc.

MANUFACTURING JOBS

Jobs at all skill levels make up the manufacturing career cluster. You might find yourself working at an auto plant assembling cars. Or you might work in a lab creating robots. Here are some common industry jobs and the skills they require.

CAREER PATHWAY ▶ Production

TOOL AND DIE MAKER

A tool and die maker produces tools, dies, and special guiding and holding devices. These products help machines mass-produce metal parts. Some of the most skilled production workers in manufacturing work in this field. Dies can be used to create everything from washing machine parts to CDs. Tool and die makers are employed in nearly all areas of manufacturing. They work in the automotive, aircraft, construction, and farm-machinery industries.

Special Skills Accuracy and attention to detail are essential in making tools and dies. Workers in this career need mechanical ability, mathematical knowledge, and the ability to work well with their hands. They must be able to read blueprints. They must also be familiar with various production materials, including metals and alloys.

WELDER, CUTTER, AND WELDER FITTER

Welding is the joining of metal parts through the use of heat. Welders spend most of their day working with metal. Most welders work for companies that manufacture transportation equipment, industrial machinery and equipment, or metal products.

To perform or supervise welding safely, workers need to wear special shoes, goggles, hoods with protective lenses, and other safety equipment. As technology advances, an increasing number of welding tasks are performed using automation or robots.

Special Skills Welders must often weld materials that are in awkward places. For this reason, they must be able to perform tasks lying on their backs or while suspended high above the ground. Welding that is performed manually requires physical stamina. To plan their work, welders must be able to read drawings, blueprints, and specifications. They must also be able to select the right tools, devise procedures, and inspect their finished product. Welders also need skills such as good near vision, precise control, and arm-hand steadiness.

Copyright © The McGraw-Hill Companies, Inc.

Copyright © The McGraw-Hill Companies, Inc. PHOTO: Pixtal/SuperStock.

CAREER PATHWAY ▶ # Manufacturing Production Process Development

MANUFACTURING PRODUCTION TECHNICIAN

Manufacturing production technicians set up, test, and adjust machinery and equipment. Production technicians need to be able to adjust machinery to ensure that it creates a high-quality product. Technicians might use tools such as calipers, micrometers, or protractors, and ring gauges. Technicians also troubleshoot problems with equipment or products. At the end of the production process, they inspect the product to make sure it meets customers needs.

Special Skills Manufacturing production technicians need specialized knowledge in a number of areas. For example, they need to understand machines and tools thoroughly. This may include machine design, uses, repair, and maintenance. They must have good mathematical skills, as well as knowledge of design techniques. They also need good problem-solving skills in order to find effective ways to fix or adjust machines.

ROBOTICS ENGINEER

Robotics engineering is behind some of today's most dynamic and innovative manufacturing projects. Advances are constantly transforming this career.

Robotics engineers are employed in a variety of industries. Some engineers design robots for use in outer space or on the ocean floor. Others automate production lines for aircraft manufacturers. In addition to developing robots, robotics engineers maintain existing robots and develop new applications for them.

Special Skills Robotics engineers spend much of their time conducting trials and research. This career requires creativity and research skills. It sometimes takes years of research and labor to create a new robot or automated system. Engineers must have extensive training and a thorough grasp of mathematics and mechanics. They must have the patience and follow-through to work on complicated projects.

Many companies that develop robotics have laboratories that are used primarily for research and development. A lab engineer might maintain and service hardware and software. She or he might demonstrate automation and robotics capabilities. Finally, lab engineers often plan, execute, and document lab experiments.

▶ # Maintenance, Installation, & Repair

MEDICAL EQUIPMENT REPAIRER

Medical equipment repairers maintain, adjust, and repair equipment used in hospitals and other health care settings. They may work on patient monitors, x-ray machines, electric wheelchairs, and other devices. These workers use a wide range of equipment to do their work. Often they use specialized software and computers. They may also use hand tools, soldering irons, and other electrical tools.

Medical equipment repairers work during the day, but they may be on call during all hours. There may be an emergency in which a repair is urgently needed. They may need to travel long distances for a repair job. They must also be willing to work in a medical environment, which can pose health risks.

Special Skills Workers in this field must be willing to work under pressure. They should be able to work well with their hands and have good technical skills. They must be able to quickly identify a problem and find a solution. They may also need a degree in biomedical technology or engineering.

GENERAL MAINTENANCE AND REPAIR WORKER

General maintenance and repair workers take care of and fix machines, mechanical equipment, and buildings. Their responsibilities vary widely. They might do pipe fitting, welding, or carpentry. They may repair boilers, machines, or electrical or mechanical equipment. Some of their work may be as simple as unclogging drains or repairing electrical switches.

Maintenance and repair workers inspect equipment, identify problems, and decide on the best way to fix them. They use hand tools, such as screwdrivers and drills, as well as more specialized equipment.

Special Skills General maintenance and repair workers should be good with their hands and have good math skills. Since the job requires standing, lifting, and climbing, good physical condition is important. These workers should also be problem solvers who can work with little supervision.

Copyright © The McGraw-Hill Companies, Inc. PHOTO: Fotosearch/SuperStock.

Copyright © The McGraw-Hill Companies, Inc. PHOTO: Jupiterimages/Getty Images.

CAREER PATHWAY # Quality Assurance

INSPECTOR, TESTER, SORTER, SAMPLER, AND WEIGHER

These workers, also known as quality-control inspectors, inspect a variety of manufactured products, from food to clothes to cars. They make sure products meet minimum quality standards. They ensure that foods are safe to eat, cars are safe to drive, and that each product is made properly. These workers may check products by sight, sound, feel, smell, or taste. They may conduct tests to make sure mechanical products will not break down in the real world.

Quality-control inspectors work at many stages of the production line. They may test the materials that will be used to assemble a product. They may also give a final check after production is complete. They may use sensitive gauges, meters, and other tools to inspect a product.

Special Skills Quality-control inspectors should have good mechanical skills, math skills, communication skills, hand-eye coordination, and vision. They may need to be able to interpret blueprints, data, and manuals. In many cases, a high school diploma is enough to begin work in this field, and workers are trained on the job.

QUALITY CONTROL SYSTEM MANAGER

Quality control system managers are responsible for planning, directing, and coordinating quality assurance programs. They develop quality control policies and make sure the products created meet the quality control standards. They may test products themselves and also oversee workers who test products. After a test is run, they may analyze test results and share the results with other managers or staff. Quality control system managers may also work with sales and marketing to learn more about what the customer expects.

Special Skills As managers, these workers must have strong communication skills. They should be able to support their employees and motivate them to achieve at high levels. They must also be willing to take risks and make important decisions. Quality control system managers should enjoy studying and solving real-world problems.

Logistics & Inventory Control

SHIPPING, RECEIVING, AND TRAFFIC CLERK

Employees in shipping, receiving, and traffic track incoming and outgoing shipments. Their duties include preparing items for shipment. They assemble goods and address and stamp merchandise. They also unpack and verify incoming merchandise or materials. Shipping, receiving, and traffic clerks are involved on a daily basis with the transportation of products.

Important tasks include preparing documents such as work orders and shipping orders. Clerks must make sure that shipping materials are in stock. When products are damaged or missing, or do not meet specification, clerks work with company representatives to fix the problem. For their various tasks, workers use tools ranging from bar code reader equipment to hand trucks.

Special Skills Shipping, receiving, and traffic clerks need to be attentive to detail and possess good people skills such as active listening and problem sensitivity. They also need to communicate well, both verbally and in writing, with supervisors, peers, and subordinates.

CARGO AND FREIGHT AGENT

Cargo and freight agents organize the movement of cargo and freight shipments, which are goods shipped for commercial purposes. They send and receive shipments from airline, train, and trucking terminals, as well as shipping docks. Agents take orders from customers and arrange for pickup and delivery of freight. They also prepare and examine documents to determine shipping charges and tariffs, or taxes. Cargo and freight agents may also pack goods for shipping.

Special Skills Cargo and freight agents need detailed knowledge of the transportation industry—the business of moving people or goods by air, rail, sea, or road. They must have good communication skills and be able to provide excellent customer service. It is helpful for agents to know one or more foreign languages, as well as have a good command of English. Agents should have good critical thinking skills, using logic and reasoning to identify the strengths and weaknesses of possible solutions to problems.

Copyright © The McGraw-Hill Companies, Inc. PHOTO: Alistair Berg/Digital Vision/Getty Images.

CAREER PATHWAY # Health, Safety, & Environmental Assurance

INDUSTRIAL SAFETY AND HEALTH ENGINEER

Safety and health engineers plan, carry out, and coordinate safety programs in the manufacturing industry. They use engineering principles and technology to prevent or correct unsafe workplace conditions. These employees often inspect machinery to identify potential hazards. They install safety devices on machines. They report or review findings from accident investigations. They also review employee safety programs to make sure workers are safe.

Special Skills Safety and health engineers need a detailed knowledge of specific areas of law, chemistry, engineering, and design. They also need effective communications skills. Finally, they need problem sensitivity. This is the ability to tell whether something is wrong or is likely to go wrong.

ENVIRONMENTAL COMPLIANCE INSPECTOR

Compliance inspectors investigate sources of pollution to protect the public and the environment. They ensure that companies follow federal, state, and local regulations. They determine code violations. They also examine permits, licenses, applications, and records to ensure that companies meet licensing requirements.

Special Skills Environmental compliance inspectors often need to work with specialized tools, including hydrocarbon analyzers, sampling pumps, and soil core sampling apparatus. They need familiarity with analytical or scientific software. They also need an extensive knowledge of laws, court procedures, and government regulations.

ENVIRONMENTAL ENGINEER

Environmental engineers look for solutions to environmental problems. They might work to solve air pollution, recycling, waste disposal, or public health issues. These engineers study hazardous waste, its effect on the public, and how to lessen the problem. They may design water supply and wastewater treatment systems. Many work as consultants, helping clients comply with, or adapt to, regulations and minimize damage to the environment.

Special Skills An environmental engineer should have a bachelor's degree or higher in the field. A wide range of math and science skills is essential. Environmental engineers should also be creative, analytical, and detail-oriented.

Copyright © The McGraw-Hill Companies, Inc.

CHAPTER

3

BUILDING A CAREER IN MANUFACTURING

Once you have found a field that interests you, look ahead and consider your career path. This path is made up of the job experiences and career moves that lead you toward your career goal. You may take several steps before reaching your ultimate goal. You will likely spend time in an entry-level position. This will help you gain the professional experience necessary to move ahead in your career.

For example, a robotics engineer can begin his or her career as an apprentice. After a few years, this employee's responsibility might increase to include designing and managing projects. Robotics engineers with significant experience generally become more involved with clients and upper-level executives. Some senior-level engineers might even start their own companies.

Don't worry if you change your mind about your career path. This happens to many people. It often takes time to find the right path. You can always change your career path regardless of where you are in your chosen profession.

Evaluating Career Choices

Choosing a career is challenging. Now is a good time to start thinking about what kind of career path you would like to follow. A well-chosen career can bring satisfaction and success in life.

Self-knowledge is the key to making wise career choices. Friends, teachers, and family members may offer helpful suggestions for potential careers. However, you are ultimately in charge of making your own career decisions.

Consider your personality, interests, aptitudes, and values when choosing a career. Think about why you chose to read this book in the first place. Of the many industries in which you could work, why does manufacturing appeal to you?

You might feel that your personality—the way you think and behave—is well suited to this industry. If you are practical, realistic, and interested in producing a product you can see and touch, then manufacturing may be a good career choice. Many workers in the manufacturing industry get personal satisfaction from creating a final product.

Copyright © The McGraw-Hill Companies, Inc.

Copyright © The McGraw-Hill Companies, Inc. PHOTO: Martin Barraud/OJO Images/Getty Images.

You should also allow your interests to influence your career decisions. What activities do you enjoy? What classes did you enjoy in school? People who like working in manufacturing typically prefer math and science classes over English and social studies classes.

In some cases, your aptitude, or ability in a certain area, will shape your career goals. Ask yourself what skills come naturally to you. A strong aptitude in math might help you become a successful machinist or civil engineer. It's important to recognize, though, that a strong aptitude is not the same as a strong interest. You might have very strong artistic abilities, for example, but prefer a career that uses other skills.

Values are another factor to consider when selecting a career. Values are the principles and beliefs that you live by. You might value responsibility, compassion, courage, recognition, independence, or creativity. Your values will shape all areas of your life, from your long-term goals to the lifestyle you lead. For example, if you place a high value on family time, you might seek a job that allows for flextime and provides good vacation benefits.

Working with Data, People, and Things

Most careers offer opportunities to work with a combination of data, people, and things. Working with data involves the evaluation of information. A job that focuses on people will be based on human relationships. Working with things involves using objects, such as tools, objects, and machines. Most jobs focus mainly on one of these. Environmental engineers, for example, work mainly with data. Labor relations managers work primarily with people. Welders work mostly with things. When planning your career path, consider what balance of data, people, and things you want in a career.

CAREERS THAT INVOLVE WORKING WITH DATA

Working with data means working with words, ideas, concepts, and numbers. Examples of working with data include preparing financial statements and drawing up budgets, making measurements and calculations, and scheduling the steps needed to manufacture a product.

Many manufacturing jobs focus on data. Machinists and tool and die makers review blueprints before calculating where and how to cut into something. Engineers may be called upon to design new products or improve existing ones. This requires conducting research, running tests, and collecting and analyzing data. Engineers may also perform calculations and make projections based on collected data.

Some managers in manufacturing also need to work with data. A production manager, for instance, may need to study spreadsheets that display rates of production to determine if a particular procedure is efficient and cost-effective.

Are you good with words and numbers? Do you enjoy applying scientific and mathematical principles to everyday situations? Do you grasp new concepts quickly? Do people say that you think logically? If so, you may want to consider a career that focuses on working with data.

CAREERS THAT INVOLVE WORKING WITH PEOPLE

Some manufacturing jobs focus on human relationships. Examples of working with people include training employees, mediating conflicts among coworkers, negotiating prices with suppliers, and advising customers. All of these activities involve strong communication skills.

Many manufacturing jobs enable you to spend a great deal of time working with others. Production managers or supervisors must effectively supervise and lead teams of workers. Assemblers rely heavily on teamwork, which requires the ability to follow directions and communicate ideas to other members.

People who enjoy working with other people are generally outgoing. To decide if you're a "people person," ask yourself a few questions. Do you place great emphasis on your friendships? Do you spend your spare time socializing with friends or family? Are you good at judging the motivations and feelings of others? If so, you'll probably enjoy a job that allows for frequent interaction with others.

Copyright © The McGraw-Hill Companies, Inc. PHOTO: Chris Ryan/OJO Images/Getty Images.

Copyright © The McGraw-Hill Companies, Inc. PHOTO: Glowimages/Getty Images.

CAREERS THAT INVOLVE WORKING WITH THINGS

In manufacturing, working with things involves designing, creating, using, and repairing machines, tools, instruments, and products. Examples of working with things include setting up and operating machinery, driving a forklift, and welding metal.

For instance, someone who works with machinery might start, stop, and observe the operations and actions of equipment. A tender (a worker who attends to machines) may need to adjust materials or controls of a machine. He or she may change guides, adjust timers and temperature gauges, turn valves to allow the flow of materials, and flip switches in response to lights.

Setting up requires preparing machines or equipment for operation by planning the order of the steps. Employees with this focus adjust the positions of parts or materials, set controls, and verify the accuracy of machine capabilities. These employees evaluate the properties of materials. They use tools, equipment, and work aids such as precision gauges and measuring instruments. This job requires experience and good independent judgment.

Most careers in manufacturing involve working primarily with things. For example, welders, assemblers, and machinists spend a majority of their day working with tools, machines, and other equipment.

Working with things, though, may occur in a wide variety of settings. For instance, boiler tenders are responsible for handling and monitoring large equipment. Boiler tenders work in a variety of manufacturing industries. Some, for example, are employed at plants that produce lumber, paper, chemicals, iron, or steel. Others work at public facilities, such as schools or government buildings.

Think about how you choose to spend your spare time. Do you enjoy building, assembling, or repairing things? Are you curious about how machines like cars or computers actually work? Do you prefer to work with your hands? If so, you're probably well suited for working with things.

Data, People, and Things in the Manufacturing Industry

Whatever career in manufacturing you choose, you're likely to spend some time working with data, people, and things. Choosing a job that matches what you like to do will make you a better employee and a happier person.

How can you find a job that best suits your strengths? One way is by browsing the Dictionary of Occupational Titles (www.occupationalinfo.org). This resource lists a wide range of jobs. Each job has a nine-digit code that identifies and describes it. The fourth, fifth, and sixth digits show how much each job involves working with data, people, and things, respectively. The lower the number, the more complex the particular type of work. For example, the code for tool and die makers is 601.260-010. The digits 260 mean complex work with data, less complex work with people, and very complex work with things.

Finding Employment

Finding a job is seldom easy, but finding one in a new career field can be even harder. Whether you have a job and are considering a career change or are unemployed, now is a good time to explore new careers and make yourself more attractive to employers.

CHANGING CAREERS

Many people jump from one career right into another. They may feel that their job does not match their skills or interests. They may believe the job does not offer enough room to advance. A new career can offer different opportunities.

The best time to think about a new career is when you are already employed. While you have a job, there is less pressure to find a new job right away. Investigate which career fields have good opportunities in the area where you live. Think about your current job. What aspects of it do you enjoy? Which other careers involve similar tasks?

If you find a job that you would like to pursue, spend time investigating the qualifications required. You might also speak to someone who works in the industry. Learn as much as you can to ensure the career cluster is right for you.

Copyright © The McGraw-Hill Companies, Inc. PHOTO: Monty Rakusen/Cultura/Getty Images.

Look for ways you can gain experience that will help you in your search. If the new career involves working with people, volunteer for tasks in which you will interact with people. You should also spend time creating a résumé. Use print and online resources to learn how to create the résumé that best highlights your qualifications. Highlight the skills that are most relevant to the jobs you will apply for.

You should also spend time networking, or reaching out to people who can help in your job search. This may include family, friends, or colleagues from current or former jobs. Make an effort to meet new people in order to expand your network. One good way to do this is to use online networking sites.

UNEMPLOYMENT

Being unemployed can be a difficult time, but it also brings new opportunities. Millions of people are unemployed at any time, so there is no shame in being unemployed. If you find yourself unemployed, be sure to apply for unemployment benefits. Benefits are only given once you are approved, so be sure to apply right away.

Make the most of your time while you are unemployed. Work on your résumé. Expand your network. Try to stick to a daily schedule. For example, you might shower and dress as if going to work, then spend the morning crafting your résumé or searching for jobs. Rather than spending every waking hour looking for work, set aside some time for leisure activities.

Consider finding a freelance or part-time job that can help you gain new skills and earn more money while you search. You might also take a class that teaches skills useful in a new career.

Copyright © The McGraw-Hill Companies, Inc. PHOTO: Monkey Business Images/Cutcaster.

EDUCATION AND TRAINING

Jobs in the manufacturing industry require varying levels of education. Many jobs in the industry require little or no formal training. For others, it is necessary to have specific education and experience.

Training and Education for Manufacturing

The level of training necessary to succeed in the manufacturing industry varies by career. Jobs can be categorized into three groups—jobs requiring little or no training, jobs requiring some training, and jobs requiring advanced training.

JOBS REQUIRING LITTLE OR NO TRAINING

Positions such as precision assembler, machine tool operator, and hand packer usually require little previous training. The training for jobs such as these, and for similar on-the-floor production jobs, is usually provided by your employer when you are hired.

However, getting a high school education or the equivalent is important. Many employers look for workers who earned solid grades in high school. Failing to complete high school may hurt your chances for advancing in your field.

JOBS REQUIRING SOME TRAINING

Many jobs in manufacturing require some specialized training. If you're interested in working as a welder, electrician, machinist, or mechanical drafter, you'll need to pursue specialized degrees or certification.

You can complete apprenticeships to prepare for many of the more skilled trades in manufacturing. Technical colleges offer programs in such fields as electronics servicing, mechanical drafting, and electronic engineering technology. Some programs will provide you with an associate's degree upon completion.

Copyright © The McGraw-Hill Companies, Inc. PHOTO: Jose Luis Pelaez Inc/Blend Images/Getty Images.

Trade organizations, or organizations representing a specific industry or type of job, often offer certification programs. These programs provide training for a specific occupation. Specialized certification exists in a wide variety of manufacturing careers, including occupations in plastics and welding. These programs might require a minimum number of hours on the job and a minimum score on a standardized test.

Earning an associate's degree or a nationally recognized certification can help you prosper in your field. For more information about such programs, contact trade organizations or community colleges and technical schools in your area. You can also learn more about trade organizations by visiting the websites listed on page 38.

JOBS REQUIRING ADVANCED TRAINING

Jobs in manufacturing that require extensive training usually involve business management, science, or engineering. Such jobs generally require at least a bachelor's degree. Some may require study at the master's or doctoral level.

Manufacturing professionals who work in scientific or engineering positions often need extensive schooling. The manufacturing industry has a wide variety of scientific and engineering jobs.

For example, the aerospace industry employs a large number of physicists. Most physicists have doctoral degrees in physics. Other manufacturing fields, such as drug manufacturers, employ chemists with doctoral degrees in chemistry.

TRAINING REQUIRED FOR MANUFACTURING JOBS

Jobs Requiring Little or No Training

Boiler Tender	Machine Tool Operator
Cooling and Freezing Equipment Operator	Shipping Clerk

Jobs Requiring Some Training or Education

Boilermaker	Patternmaker
Computer Technician	Quality Control Technician
Machinist	Tool and Die Maker
Millwright	

Jobs Requiring Advanced Training or Education

Computer Engineer	Mechanical Engineer
Industrial Engineer	Robotics Engineer

Copyright © The McGraw-Hill Companies, Inc.

The manufacturing industry also employs many engineers, particularly in the automotive and aerospace industries. To become an engineer, you must have at least a bachelor's degree in engineering. Common types of engineers in manufacturing include mechanical, chemical, electrical, and civil engineers. Many engineers eventually pursue advanced degrees (master's or doctoral degrees) in their fields.

Pre-Employment Training

Pre-employment training for positions in manufacturing may involve completing one or more of the following:

- an apprenticeship or internship

- a certification or specialized program at a technical school

- a bachelor's or master's degree program at a college or university

Before seeking formal training, look for ways to get on-the-job training. Some employers provide such training for employees in entry-level jobs.

APPRENTICESHIPS

An apprenticeship is a way to gain real-life work experience. In an apprenticeship, an inexperienced worker learns a trade by working alongside an expert employee. Some apprenticeships last as long as four to five years. The apprentice may earn little pay. However, workers who have completed these programs are often well respected and well paid. Some unions and professional associations in manufacturing sponsor apprenticeships. An excellent source of leads for apprenticeship opportunities is the US Department of Labor's Office of Apprenticeship (http://oa.doleta.gov).

INTERNSHIPS

An internship is an opportunity to gain practical experience in a field. Internships are usually shorter than apprenticeships. They may offer the chance to learn about various departments in a company. An intern often receives little or no pay. However, completing an internship can improve your chances of getting a job.

Copyright © The McGraw-Hill Companies, Inc. PHOTO: Stockbroker/SuperStock.

Many engineers and other manufacturing professionals complete internships before entering the workforce. Many companies post information about internship opportunities on their websites. Job seekers interested in such opportunities can find out more by calling or e-mailing the appropriate individuals listed on the sites.

TECHNICAL SCHOOLS

If you're interested in a position that requires training, a technical school is a promising option. A technical school offers skills-oriented programs. Examples include computer repair or data processing. The occupational areas most commonly taught in technical schools are business, skilled trades (such as automotive repair and carpentry), health care, agriculture, family and consumer services, and information technology.

POSTSECONDARY EDUCATION

Most skilled and professional workers in manufacturing have completed some form of postsecondary education (study conducted after high school). Most people seek postsecondary education at universities, technical schools, or community colleges.

Upper-level and management workers often have undergraduate or graduate degrees in business and engineering. Skilled production workers may have special training in technology such as computer-aided drawing (CAD).

Several colleges offer programs to train professionals in the manufacturing industry. Programs include training in tool and die technology, sheet metal technology, boilermaking, manufacturing engineering, pharmacology, and textile science. These specialized programs often provide access to internships and jobs in the industry. Through these programs, you may meet people in the industry who could later help you get a job.

Before choosing a postsecondary educational program, make sure that it will prepare you for the job you want. Consider the length of the program and its rate of job placement. Also check to see that the program is nationally accredited. Take into consideration a school's reputation in the field and the expertise of its faculty.

Another important factor is cost. You may be able to finance your education through loans, grants, work-study programs, or scholarships. Contact the financial aid office of the schools that interest you to find out which scholarships you might qualify for.

Copyright © The McGraw-Hill Companies, Inc.

POSTGRADUATE EDUCATION

Postgraduate education is conducted after the completion of a bachelor's degree. Most people who seek postgraduate education enroll in master's or doctoral programs.

In the manufacturing industry, many upper-level scientists, researchers, and engineers have completed a postgraduate education. Production managers and other management personnel may also have earned advanced degrees. As manufacturing becomes more high-tech, an increasing number of firms prefer job candidates who have completed at least a four-year degree. A master's or doctoral degree may give a candidate an advantage.

Industries such as aerospace, pharmaceuticals, and chemical manufacturing have traditionally employed a large number of professionals with advanced degrees. The complicated nature of the work and the research demanded of employees in these industries makes advanced study a necessity.

Individuals who want to work as scientists, researchers, engineers, and upper-level managers should plan on seeking postgraduate education. Many professionals, however, put off graduate school. Instead, they take an entry-level job in their field. This enables them to gain experience. They may qualify for company-sponsored tuition-assistance programs.

When universities evaluate applicants for master's and doctoral degree programs, they consider a candidate's academic record and sometimes his or her work experience. In most cases, admission is very selective. Those who wish to earn postgraduate degrees must have a strong academic background.

On-the-Job Training

On-the-job training is on-site instruction in how to perform a particular job. If you are seeking a position that requires little training, your employer will most likely train you.

On-the-job training offers several benefits. First, this type of training is usually paid. Even if it is not paid, you will still not need to pay for the knowledge and skills you'll gain. Another benefit is that the training is tailored to the job. By the time you complete your training, you'll probably feel comfortable in your new position.

Copyright © The McGraw-Hill Companies, Inc. PHOTO: Comstock Images/Getty Images.

TYPES OF ON-THE-JOB TRAINING

Many manufacturing jobs offer on-the-job training. On-the-job training is especially common for jobs where workers are in high demand. For example, there is an increased need for workers who can use laser technology. The supply of these workers has not kept pace with the demand. Many employers have chosen to provide on-the-job training in this skill.

In some cases, even high-level jobs provide on-the-job training. For example, many companies have management-training programs in production. There is a great deal to learn in the field of production management. Therefore, it may take years for a trainee to become a full-fledged production manager.

Many paid apprenticeships also qualify as on-the-job training, especially if they are sponsored by the company employing the apprentice. Such apprenticeships may require an apprentice to work at the sponsor company for a prescribed length of time.

Job and Workplace Skills

When considering job candidates, employers look for both job-specific skills and general workplace skills. Job-specific skills are the skills necessary to do a particular job. They may include balancing a budget or programming a computer. General workplace skills can be used in a variety of jobs.

MANUFACTURING SKILL STANDARDS

The Manufacturing Skill Standards Council (MSSC) represents individuals and organizations from the manufacturing industry. These include labor unions, companies, educators, and public interest groups. They have worked together to create a set of core knowledge and skills. The standards stress the importance of specific job skills. They also emphasize general workplace skills. The standards call for knowledge in 17 academic areas, such as math, science, computer technology, and writing. The standards also define three main technical areas: safety procedures, the manufacturing process, and business policies and procedures.

The MSSC has developed skill standards for jobs in the six manufacturing career paths. These are production; manufacturing production process development; maintenance, installation, and repair; quality assurance; logistics and inventory control; and health, safety, and environmental assurance. (See Chapter 1.) These standards outline how work should be performed. The standards also define the level of knowledge and skill required for specific careers. You can learn more about the standards at www.msscusa.org.

Copyright © The McGraw-Hill Companies, Inc.

CORE SKILLS

Core skills differ from academic or job-specific skills. They are learned both inside and outside the classroom. They are transferable from job to job. Developing these skills will make you more marketable in manufacturing jobs, and in any job situation.

Communication Skills In manufacturing, workers need good communication skills when offering instruction to the people they manage or when bringing problems to management's attention. Good communication is also needed to develop good, productive relationships with coworkers.

Listening Skills Listening is the foundation of learning. Active listening skills are vital to effective communication. Listening skills are essential for following instructions safely and precisely. In the manufacturing industry, this is necessary to ensure a safe working environment. Good listening skills also help colleagues understand each other's ideas and points of view.

Problem-Solving Skills Employers value workers who can spot problems and take action to find solutions. Solving problems requires creativity and self-reliance. A production supervisor, for example, may need to solve complex problems occurring on the production line or between departments. Maintenance, installation, and repair workers confront a broad range of problems almost every day. They have to identify problems and then evaluate potential solutions. The same is true for manufacturing employees in the logistics and quality assurance career paths.

Technology Skills Technology has an ever-increasing impact on production processes. Many plants now use computer automation instead of human workers to produce items more quickly and efficiently. Line workers, such as assemblers, may be required to run computer programs and operate high-tech machinery.

Copyright © The McGraw-Hill Companies, Inc. PHOTO: Monty Rakusen/Cultura/Getty Images.

Decision-Making Skills Time is of the essence in manufacturing. The ability to gather and analyze information rapidly, and to think clearly under pressure, is key to this industry. A boiler tender, for example, might be faced with an equipment breakdown or problem during a night shift that will require him or her to make effective decisions quickly. Employees in the health, safety, and environmental assurance pathway must make important decisions daily to ensure a safe work environment.

Organizing and Planning Skills Because manufacturing involves so many steps, organization is crucial. Planning requires the ability to set goals and to visualize the sequence of steps leading up to those goals. Robotics engineers, for example, might be called on to develop new products on a deadline. They would then have to make a plan and organize different departments, teams, and processes to make sure the deadline is met.

Teamwork Skills Teamwork is key in manufacturing. For example, a successful manufacturing firm brings hundreds, perhaps even thousands, of products to market. To do so, producers must work with maintenance and repair employees, logisticians, quality assurance workers, and health and safety inspectors. Good communication is one of the most important foundations of teamwork.

Social Skills Some jobs may not require much interaction with coworkers. However, respectful interaction with coworkers makes for a more enjoyable work atmosphere. In a production factory, workers might interact with individuals from various levels of the company, from apprentices to managers. They should keep an open mind and use the opportunity to learn from others. Some studies have shown that between 80 to 85 percent of a person's success in the workplace is due to the person's social skills. These skills build an excellent foundation for advancement opportunities.

Adaptability Skills In manufacturing, job descriptions, work environments, and production processes are constantly changing because of technological innovation. Workers should keep an open mind and be ready to acquire new skills. Remember that for every firm in the manufacturing industry, the image of being on the cutting edge is highly desirable.

Copyright © The McGraw-Hill Companies, Inc. PHOTO: Fuse/Getty Images.

WORKING IN THE MANUFACTURING INDUSTRY

When choosing a career path, it is important to know what it is like to work in the industry. Understanding the work environment, hazards, and benefits of a job can help you make informed decisions.

Work Environment

Work environment refers to factors that affect workers' health and satisfaction on the job. These include the physical surroundings and the working hours. They also include the physical activities required to perform the job.

PHYSICAL ENVIRONMENT

The physical work environment in manufacturing varies by job and by setting. Some jobs require workers to spend long days on their feet in noisy factories. Other jobs require them to handle chemicals and toxic substances. Most factory buildings are clean and have good lighting and ventilation. Some factories even have outdoor facilities.

In the aerospace, apparel, pharmaceutical, electronics, automotive, textile, and computer industries, employees usually enjoy a reasonably pleasant work environment. In every industry, however, production workers confront greater environmental stresses than administrators and engineers do. Workers in the iron and steel industries, for example, are often exposed to intense heat and noise. In glassmaking, which requires the use of high temperatures and heavy machinery, there is the risk of cuts and burns.

Copyright © The McGraw-Hill Companies, Inc. PHOTO: DreamPictures/Shannon Faulk/Getty Images.

WORK HOURS

The work hours of jobs in manufacturing vary greatly. Because the equipment in many manufacturing plants requires 24-hour supervision, shift work is vital in this industry. Shift work divides the day into blocks of time, generally eight hours. Shift work allows manufacturing companies to operate around the clock. It also allows workers to select hours to meet their needs. Shift work is common for assemblers, boiler tenders, quality control inspectors, and production supervisors.

WORKING CONDITIONS IN MANUFACTURING	
Production	• Workers are sometimes exposed to high noise levels, and heavy lifting may be required in some jobs. • Part-time work is unusual. • In chemical plants, split, weekend, and night shifts are common, but pay is usually higher for non-traditional hours. • Many food industry production jobs involve repetitive, physically demanding work.
Manufacturing Production Process Development	• In the computer industry, research and development (R&D) personnel may work long hours of overtime. • Many food manufacturing plants have redesigned equipment and increased the use of job rotation. • In textiles, travel is an important part of the job for many managers and designers.
Maintenance, Installation, and Repair	• In steel mills, computer-controlled machinery helps to move iron and steel through the production processes, reducing the need for heavy labor.
Quality Assurance	• In pharmaceuticals, the danger of contamination means that there is rigorous attention needed in keeping plants and equipment clean. • In food manufacturing, managers and employees must comply with numerous government standards and regulations.
Logistics & Inventory Control	• In motor vehicles manufacturing, overtime is especially common during periods of peak demand.
Health, Safety, & Environmental Assurance	• Environmental engineers and health and safety engineers spend much of their time in offices working on plans and safety reports. • Engineers may travel on-site to inspect facilities and machinery. In some cases, they may spend time near hazardous materials and situations to find ways to reduce the hazard's impact and improve safety.

Source: US Department of Labor, *Career Guide to Industries 2010–2011* and *O*NET*

Copyright © The McGraw-Hill Companies, Inc.

Flextime is a trend affecting all industries, including manufacturing. Flextime allows workers to choose the hours and days they work. However, employees must maintain a standard total number of hours per week. Workers may adjust their hours to suit their personal needs. One employee may choose to work ten hours only four days a week, for example. Another employee may work six and one-half hours six days a week. Flextime is especially important for workers who have young children or other family commitments.

Machinists, tool and die makers, and precision assemblers typically work eight-hour shifts. Although they do most of their work during standard working hours, they may need to work overtime to meet production demands. Most welding technicians work 40 hours per week, although opportunities for overtime are usually available. Boiler tenders also generally work 40 hours per week. However, because boilers operate around the clock, most boiler tenders are required to work some nights, weekends, and holidays.

Many workers in manufacturing must deal with tight deadlines. A deadline might require that a specific number of cars be produced for a certain launch date, for example. This deadline would affect production supervisors, assemblers, and line workers. Manufacturing engineers and robotics engineers often face tight deadlines when developing new and innovative products.

ESSENTIAL PHYSICAL ACTIVITIES

Production workers in manufacturing must be in good physical shape to withstand the demands of their jobs. Heavy lifting and the operation of powerful equipment are required for most production jobs. It is vital that workers be trained to handle heavy equipment properly. Although some production jobs are not strenuous, most workers are required to stand, walk, stoop, bend, or climb ladders regularly during the day.

Good vision is a must for production workers who have to test products by color. Good eyesight is also necessary for the accurate reading of thermometers, gauges, charts, and meters.

Copyright © The McGraw-Hill Companies, Inc. PHOTO: Design Pics / SuperStock.

Manufacturing workers should be skilled with their hands. Many assembly-line workers handle small pieces and parts and detailed machinery. Machinists and precision assemblers must have good hand-eye coordination to work with the machinery and components.

Hazards and Environmental Dangers

Because accidents can happen on any job, safety must be a priority. The federal government protects individual workers on the job through agencies such as the Occupational Safety and Health Administration (OSHA). They create safety standards and laws that help prevent accidents. They also ensure that accident victims are offered assistance.

INJURIES AND ILLNESSES

Most on-the-job impairments are either occupational injuries or occupational illnesses. An occupational injury is any injury that occurs at work. Injuries may include cuts, fractures, or sprains. An occupational illness is caused by on-the-job exposure to harmful substances. Illnesses may include rashes and skin diseases, respiratory problems, or poisoning.

RATES OF WORK-RELATED INJURIES AND ILLNESSES IN THE MANUFACTURING INDUSTRY PER 100 FULL-TIME WORKERS (2009)

Occupation	Rate
Wood product manufacturing	6.5
Warehousing and storage	5.9
Nonmetallic mineral product manufacturing	5.2
Transportation equipment manufacturing	5.2
Furniture and related product manufacturing	5.2
Plastics and rubber products manufacturing	4.8
Truck transportation	4.6
Machinery manufacturing	4.3
Support activities for transportation	4.0
Electrical equipment, appliance, and component manufacturing	3.5
Textile product mills	2.9
Chemical manufacturing	2.3
Computer and electronic product manufacturing	1.6

Source: US Department of Labor

Copyright © The McGraw-Hill Companies, Inc.

Many manufacturing jobs involve operating heavy equipment and machinery that can be dangerous if improperly used. Proper training and common sense are therefore necessary for tasks involving machinery. Workers must follow all safety procedures. They should take advantage of breaks in line work to rest tired muscles. They should also vary their tasks when possible. They should notify managers if they feel pain or stiffness.

Another job hazard is eye injuries. Goggles are necessary during exposure to chemicals, extremely bright lights, and sharp pieces. Workers who focus on very detailed machinery, such as tool and die makers, may suffer from eyestrain. Developments in ergonomics are helping to create a healthier work environment. Ergonomics is the study of creating and adjusting work equipment and practices to make workplaces safer and more comfortable. Ergonomically correct production lines, for example, are set at a comfortable height. Workers can maintain a healthy and appropriate stance and posture. Ergonomic workstations can be adjusted to accommodate workers of different heights.

Job Benefits

Benefits aren't just extras. They not only make your life easier and safer, but can also be worth 20 to 35 percent of your salary.

Standard job benefits usually include health insurance and paid holidays, sick time, and vacations. At most companies, new employees do not start receiving paid vacation time until they have been on the job for 90 days or more. The specific job benefits you receive will depend on several different factors. The size and type of the organization you work for is one. How many years you have been on the job is another.

At some companies, job benefits have expanded to include more than health insurance, paid vacation, and holidays, and sick leave. Some expanded benefits may include:

- dental, life, and disability insurance for the worker and his or her spouse or partner

- time off to care for sick children

- tuition assistance

- 401(k) plan, or a retirement plan in which employees invest a portion of their income while employers match the contribution up to a specific amount

- child-care assistance

Copyright © The McGraw-Hill Companies, Inc.

Labor Unions

In order to protect their interests and bargain as a group, some workers have formed unions. A union is a group of workers who unite to bargain for job improvements.

Union leaders negotiate for better wages, increased benefits, better working conditions, and other job improvements. If an agreement is not reached, the union may use its most powerful tool—a strike. A strike occurs when employees stop working in an effort to force an employer to agree to the union's terms. In most cases, unions maintain strike funds, which provide partial salaries to striking workers.

When an agreement is reached between the union and company management, the company signs a labor contract. A labor contract is a legal agreement specifying wages, work hours, working conditions, and benefits. The union members must approve the contract before it goes into effect.

UNIONS IN MANUFACTURING

The percentage of workers in unions varies significantly by industry. Fewer than 10 percent of workers in the pharmaceuticals industry, for example, belong to a union. A much larger percentage of workers in motor vehicle and equipment production are union members. The largest union in this area of the manufacturing industry is the United Auto Workers (UAW).

If you join a union, you will be asked to pay an initiation fee and dues. This money supports the strike fund and the work of the union. If you are thinking about joining a union, consider the membership cost and benefits. You should also consider what the union has accomplished on behalf of its members in the past. In some occupations, you may be required to join a union as a condition of employment.

Copyright © The McGraw-Hill Companies, Inc. PHOTO: Andrew Resek/The McGraw-Hill Companies, Inc.

INDUSTRY TRENDS

The manufacturing career cluster is constantly evolving. People's needs for products change. Advances in technology affect the way business is conducted. Trends resulting from an increasingly global economy also continue to affect manufacturing.

Technology in the Manufacturing Industry

It is hard for some people to imagine how manufacturers ever operated without the benefit of today's technology. Companies that once dealt with suppliers in person or by telephone, now use e-mail and websites to make purchases. Manufacturers also use the Internet to participate in online auctions. These auctions allow companies to bid on parts, products, or services over the Internet.

Manufacturers once depended on computers only for monitoring and control purposes. Today, computers are necessary for nearly every part of manufacturing.

COMPUTER-AIDED DESIGN AND MANUFACTURING

Computer-aided design (CAD) allows workers to design products using computers. Tool and die makers, for example, use CAD to create drawings and plans for their products before actually making them. Computer-aided design allows workers to create exact plans for new products in a fraction of the time the task took in the past.

Once a CAD drawing has been created, computer-aided manufacturing (CAM) is used. CAM creates a program that will actually tell a machine how to create the product in the drawing. CAD and CAM are known as automated production devices.

Copyright © The McGraw-Hill Companies, Inc. PHOTO: Monty Rakusen/Getty Images.

An outgrowth of CAD and CAM is computer-aided engineering (CAE). CAE is the process of analyzing a product or process and using computers to design detailed drawings to make improvements. CAE applies the techniques of computers and computer graphics to engineering. CAE can be used to redesign and produce lightweight plastic containers for the food and personal care industries.

ROBOTICS

Robotics is the technology used in designing, constructing, and operating robots. It is perhaps the most useful development in computer-aided manufacturing. Creating and programming robots to perform production tasks is an example of automation. Automation is the operation and control of machinery by electronic devices. Robots can be programmed to do a wide range of jobs, such as spot-welding parts of an automobile frame.

The automotive industry is one of the biggest users of robots. However, other areas of manufacturing are introducing robots into production processes as well. These include the aerospace, electronics, food processing, and pharmaceutical industries. For example, robotic devices have been used by NASA to collect and analyze soil samples on distant planets. The International Federation of Robotics predicts that worldwide sales of industrial robots may near the 100,000 mark in 2013. Some critics argue that robots are taking jobs away from people. However, robots are essential in performing jobs that are too dangerous for people to do.

JUST-IN-TIME PRODUCTION

Another benefit of technology is a just-in-time (JIT) inventory system. This system allows parts and raw materials to be delivered to production plants just before they are needed. A computer keeps track of supplies. Orders are placed when necessary items run low in stock. JIT decreases the inventory that a plant must maintain. This, in, turn, saves money. JIT also cuts down on overstock, or having too much inventory. Successful just-in-time manufacturing depends on frequent communication between manufacturer and supplier.

Computer technology is also changing manufacturing strategies. The traditional manufacturing strategy, known as "push manufacturing," is to produce as much inventory as possible. Plants using this strategy often create an oversupply of products. Today, many companies are switching to "pull manufacturing." In this strategy, products are created in response to consumer demand. "Pull manufacturing" offers better responsiveness to special orders.

Copyright © The McGraw-Hill Companies, Inc.

Contemporary Issues in Manufacturing

Manufacturers are always searching for the best way to get the job done. New methods are constantly being tested to produce goods that use environmentally safe processes. In addition, manufacturers want to reduce waste. They are also discovering that outsourcing, or sending jobs to other companies, often overseas, helps to improve production in a variety of ways.

GREEN MANUFACTURING

Green manufacturing emphasizes conserving resources and minimizing pollution and waste. The manufacturing industry generates more waste than any other industry. A growing number of environmental laws, as well as increasing consumer demand for environmentally sound products and processes, are creating new challenges for many companies.

The ultimate goal of green manufacturing is to create no waste at all. In some cases, remanufacturing can accomplish this. *Remanufacturing* is the process of taking apart, cleaning, repairing, and putting back together products for reuse. Instead of going to landfills or having their parts melted down for raw materials, products like vending machines and photocopiers are rebuilt and resold.

OUTSOURCING IN MANUFACTURING

Outsourcing is the process of turning over control of certain tasks or duties to other companies. The contracting company specifies the desired end result but does not specify how the result should be achieved. For example, an automobile factory might outsource the production of axles. The factory does not specify which machines, part numbers, or brands the outsource company should use. Instead, the contractor depends on the outsource company to complete the job in the most cost-effective and efficient manner. Many large electronics manufacturers practice outsourcing. Companies must do a great deal of research before choosing the right outsource firm.

Companies that outsource must maintain a close relationship with the outsource company. The two firms must work toward a common goal. Outsourcing relationships work best when the partner companies share similar values, goals, and expectations.

Copyright © The McGraw-Hill Companies, inc.

Copyright © The McGraw-Hill Companies, Inc. PHOTO: Steve Hix/Somos Images/Fuse/Getty Images.

Workplace Trends in Manufacturing

The manufacturing industry is affected by the changes taking place in the US workforce. In particular, the global economy, an increase in the hiring of women and ethnic minorities, and a move toward team manufacturing are major trends today.

THE GLOBAL ECONOMY

The global economy is the worldwide linking of national economies. The global economy makes free trade possible. More and more companies are opening manufacturing plants in other countries. Firms may save on labor and overhead costs this way. Lower costs mean higher profits, higher stock prices, and lower costs for consumers.

However, many people believe that the move of manufacturing jobs has cost many American manufacturing workers their jobs. In addition, some people criticize the global economy for putting the interests of large corporations ahead of the interests of workers and local communities.

DIVERSITY IN THE WORKPLACE

The manufacturing industry continues to work to create more diverse workplaces. Companies are employing more women and people of minority ethnic backgrounds. Many companies have established affinity groups. Affinity groups are forums for employees of similar backgrounds to share ideas.

Sensitivity to issues of diversity is also increasing in manufacturing. Some companies require their employees to go through intensive diversity training.

TEAM MANUFACTURING

Many manufacturing firms now aim to create a team atmosphere. Team manufacturing requires employees at a variety of levels to interact with one another. For instance, production managers might work together to develop more efficient assembly-line procedures. Engineers might collaborate with sales representatives to create marketing strategies for various products.

CAREER RESOURCES

GENERAL CAREER RESOURCES

Career Key
www.careerkey.org
A free online self-assessment test that identifies students' Holland career choice personality type.

CTE–Career Technical Education
www.careertech.org/career-clusters/glance/at-a-glance.html
A site featuring definitions and models of career clusters, along with resources about programs of study and real-world examples.

Dictionary of Occupational Titles
www.occupationalinfo.org
A searchable database of job titles and descriptions.

Mapping Your Future
www.mappingyourfuture.org
Career and education planning information for students, from middle school to adult.

Mind Tools
www.mindtools.com
A resource for developing the essential skills and techniques that will help workers excel in any chosen profession.

O*NET
http://online.onetcenter.org
This online resource center offers skills profiles, details about hundreds of individual occupations, and crosswalk to DOT codes.

Occupational Outlook Handbook
www.bls.gov/oco/
The full text of the *Occupational Outlook Handbook* online provides information on education needs, earnings, prospects, descriptions, and conditions of hundreds of jobs.

Salary.com
www.salary.com
A nationwide database of salary information for hundreds of careers.

MANUFACTURING RESOURCES

American Institute of Chemical Engineers
www.aiche.org
A professional association dedicated to advancing the chemical engineering profession.

American Iron and Steel Institute
www.steel.org
A trade organization created to promote steel and to enhance the competitiveness of the North American steel industry.

American Welding Society
www.aws.org
An association created to advance the science, technology, and application of welding and related disciplines.

The Institute of Electrical and Electronics Engineers, Inc.
www.ieee.org
A professional association for electrical and electronics engineers in areas such as computer engineering, aerospace, and consumer electronics.

Manufacturing.net
www.manufacturing.net
An online portal providing information, news, and resources about the manufacturing industry.

ManufacturingJobs.com
www.manufacturingjobs.com
A job board for job seekers and employers in the manufacturing industry.

National Tooling and Machining Association
www.ntma.org
The association represents the precision custom manufacturing industry in the United States.

Society of Automotive Engineers
www.sae.org
An association made of up engineers and related technical experts in aerospace, automotive, and commercial-vehicle industries.

Society of Manufacturing Engineers
www.sme.org
A professional society that seeks to keep manufacturing professionals up to date on trends and technologies.

Copyright © The McGraw-Hill Companies, Inc.

PREPARE

This section of *Career Companion: Manufacturing* provides practice of the skills you will need for any manufacturing career. It is divided into three workplace skill areas:

READING FOR INFORMATION

LOCATING INFORMATION

APPLIED MATHEMATICS

At the beginning of each section is a list of specific skills that will be presented. Also included are examples of situations in which these skills are likely to be used.

After practicing these workplace skills, you will be able to answer the following questions:

- How can I identify the main idea of a workplace document?
- What do I need to look for when following step-by-step instructions?
- How can workplace graphics help me make decisions?
- What types of calculations do I need to know to do my job?
- How can I solve problems using math operations?

Working your way through each skill area will help you prepare for a job in manufacturing.

READING FOR INFORMATION

Reading for information is a key skill in the manufacturing industry. You may spend your days testing equipment or overseeing workers. No matter what the job, at some point you will need to read text to gather information. Before applying for a job, you will need to read a job description and understand the duties involved. You may be required to read a job application and understand the information it asks you to provide. Once hired, you may need to read the employee handbook, which list rules and regulations for your position.

To succeed at a job, you must be able to understand the purpose of texts you encounter and identify the most important ideas and details. You must also know how to respond to them.

On the following pages, you will encounter a variety of workplace documents to read and interpret. You will also use a wide range of reading skills.

When you read a question on the following pages, think about what is being asked and how you might find the answer. Read the text carefully, focusing on the information you are asked to find or the steps you are asked to take. After you have chosen an answer, look back to make sure you have answered the question being asked.

Learning these key reading skills will speed your path to advancement in the manufacturing industry.

Copyright © The McGraw-Hill Companies, Inc. PHOTO: Dan Bannister/Blend Images/Getty Images.

KEY SKILLS FOR CAREER SUCCESS

Here are the topics and skills covered in this section and some examples of how you might use them to read different types of materials.

TOPIC	SKILL
Read and Understand Information in Workplace Documents	1. Identify Main Idea and Details 2. Identify Details That Are Not Clearly Stated

Example: As a machinery mechanic, you may need to identify the main issue in a report describing problems with a piece of equipment.

TOPIC	SKILL
Follow Instructions from Workplace Documents	3. Understand and Apply Basic and Multi-Step Instructions 4. Apply Instructions to Unique Situations

Example: As a boilermaker, you may need to follow steps in a technical manual in order to use new equipment.

TOPIC	SKILL
Define and Use Words in the Workplace	5. Determine the Meaning of New Words 6. Understand Unique Words and Acronyms 7. Understand and Apply Technical Terms and Jargon

Example: As a tool and die maker, you may need to need to find the meaning of unfamiliar words in a journal article about production industry trends.

TOPIC	SKILL
Understand and Follow Policies and Procedures in Workplace Documents	8. Apply Workplace Policies and Procedures 9. Understand the Rationale Behind Workplace Policies

Example: As a welder, you may need to read parts of the company safety manual that apply specifically to the potential hazards of your job and understand the reasoning behind the rules.

Copyright © The McGraw-Hill Companies, Inc.

SKILL

Read and
Understand
Information
in Workplace
Documents

Follow
Instructions
from Workplace
Documents

Define and Use
Words in the
Workplace

Understand
and Follow
Policies and
Procedures
in Workplace
Documents

IDENTIFY MAIN IDEA AND DETAILS

Workers in manufacturing need to be able to identify the main idea and details in documents. When reading specification manuals, quality inspectors must be able to find the main idea that explains what they should be looking for. They must also find details supporting the main idea. The main idea tells what the document is about. Details provide more information that helps explain the main idea.

SAFETY

Slips, trips, and falls are a major cause of accidents. They cause 15 percent of all accidental deaths and are second only to motor vehicles as a cause of fatalities. The walking and working surfaces within meat, poultry, and egg product plants may be hazardous.

The following are safety considerations you need to follow regarding walking and working surfaces:

- Wear skid-resistant footwear with adequate tread on the soles.
- Use the "packing-house shuffle" when walking in slippery areas.
- Walk don't run in meat, poultry, and egg product plants.
- Use all available hand and stair rails.

1. You have just been hired to work in a meatpacking plant and are asked to read a poster titled "SAFETY." What is the main idea of the poster?

   A. Slips, trips, and falls cause 15 percent of all accidental deaths.

   B. Motor vehicles are the primary cause of accidental deaths.

   C. Meat, poultry, and egg plants are very dangerous if appropriate safety precautions are not taken.

   D. Specific safety standards for walking and working surfaces must be followed in this plant.

   E. Employees must use the "packing-house shuffle" when walking in slippery areas.

2. According to the information on the poster, what clothing is appropriate for your workplace?

   A. warm, loose-fitting clothing

   B. warm, tight-fitting clothing

   C. layered clothing

   D. skid-resistant shoes

   E. leather-soled shoes

Copyright © The McGraw-Hill Companies, Inc. TEXT: U.S. Department of Agriculture, Food and Safety Inspection

POSITION: FACTORY WAREHOUSE LABORER

Job Type: Regular, Full-Time

Job Location: Warehouse in Flint, Michigan

Job Category: Manufacturing/Production/Operations

Job Duties:
- Package products for shipment.
- Ensure the correct quantity of product enters its final packaging.
- Properly seal and label all packaging for shipment.
- Ensure that packaging is clean and not damaged.
- Stack finished products on pallets.

Job Skills:
- General warehouse

Compensation:
- $9 per hour
- After 90 days employees will be evaluated for a pay increase.

Hours/Shifts:
- Second or Third

3. You are a laborer looking for job in Flint, Michigan. You see this item in a local newspaper. What is the main idea of this posting?

   A. A company wants to hire a laborer for a warehouse job.

   B. A factory in Flint, Michigan, has a warehouse.

   C. Applicants must be willing to work second and third shifts.

   D. Applicants will be eligible for a wage increase after 90 days.

   E. Jobs for warehouse laborers are available throughout Michigan.

4. Which detail in the document suggests that the company expects the worker to be a permanent employee rather than a temporary employee?

   A. The job type is regular, full-time.

   B. The job category is manufacturing/production/operations.

   C. Several job duties are listed.

   D. General warehouse job skills are required.

   E. A specific pay rate is listed.

ANSWER KEY

Item 1: **D** Specific safety standards for walking and working surfaces must be followed in this plant.

Item 2: **D** skid-resistant shoes

Item 3: **A** A company wants to hire a laborer for a warehouse job.

Item 4: **A** The job type is regular, full-time.

Copyright © The McGraw-Hill Companies, Inc.

IDENTIFY DETAILS THAT ARE NOT CLEARLY STATED

The details in workplace documents are not always clearly stated. For example, a help wanted ad for a machine repairer might ask for related experience. People applying need to identify what would qualify as related experience and determine if they have that experience. It may sometimes be necessary to infer, or make a logical guess, when a detail is suggested rather than stated.

MEMORANDUM

To: Repair Crew

From: Manager, Inventory

Re: Ordering Parts

Many of you have asked about ordering parts for the presses and other equipment at the plant. When you place an order, please make sure to say that all orders should be delivered to the attention of the inventory department. We must keep track of all deliveries and invoices that arrive at the printing plant, so we can send that information to accounting. You can pick up any parts you have ordered that are less than 50 lbs. (pounds) at the inventory department's main counter. You must pick up any parts heavier than 50 lbs. at the loading dock.

1. You are a machine repairer at a printing plant. You read the memo to determine how to order parts for a printing press that is not working. You have determined which parts you need to order and are completing your company's purchase order form. Where should you have the parts company send the parts?

   A. to your supervisor at the printing plant

   B. to the printing plant's inventory department

   C. to the printing plant's president

   D. to you

   E. to the printing plant's accounting department

2. You are in the process of repairing a binder machine, which is a piece of equipment that attaches the pages of a book together. You need to order a part for the binder that weighs 123 pounds. Where will you pick up the part?

   A. at the printing plant's loading dock

   B. at the binder machine

   C. at the main desk of the inventory department

   D. at the printing plant's main office

   E. at the printing plant's accounting department

Copyright © The McGraw-Hill Companies, Inc.

WAREHOUSE AND OPERATIONS SPECIALIST

Description

Responsibilities include: Keeping accurate records on warehouse inventory; truck loading and unloading (manually and using a forklift), and primary shipping/receiving. Forklift driving certificate may be required.

Primary Responsibilities

- Operate equipment and/or a forklift safely and efficiently
- Track, receive, and store incoming items efficiently using established procedures and automated devices including barcode scanners
- Maintain accurate, complete records of incoming, outgoing, and stored product including partial cases of returned product

Desired Experience

- Experience in a high-speed warehouse or shipping environment
- Forklift driving experience or certificate
- Tracking and managing inventory
- Computer experience including using spreadsheets, proprietary applications, and hand-held devices including barcode scanners

3. As a warehouse and operations specialist, you are interested in applying for this job. Which of the following jobs listed on your resume should you highlight because it relates directly to the type of experience the employer is looking for?

   A. restaurant busser; cleaned tables and refilled beverages

   B. theme park operator; operated a roller coaster

   C. cafeteria cashier; rang up customers' orders

   D. stock clerk; used a barcode scanner to check inventory level

   E. fry cook; managed the fryer

4. You also have additional skills and achievements listed on your resume. According to the job description's list of desired experience, which of the following should you highlight on your resume?

   A. your commercial driver's license

   B. your experience as a telemarketer

   C. your experience as a clothing store sales clerk

   D. your certificate to teach cooking classes

   E. your Employee of the Month award at a fast-food restaurant

ANSWER KEY

Item 1: **B** to the printing plant's inventory department

Item 2: **A** at the printing plant's loading dock

Item 3: **D** stock clerk; used a barcode scanner to check inventory level

Item 4: **A** your commercial driver's license

Copyright © The McGraw-Hill Companies, Inc.

UNDERSTAND AND APPLY BASIC AND MULTI-STEP INSTRUCTIONS

It may be necessary to follow multi-step instructions in a variety of situations. A shipping clerk must be able to gather, pack, and ship a variety of items for delivery according to written instructions. Workers must read carefully to know when to take each step and be able to apply the same instructions in a variety of situations.

PROCEDURE: LIPSTICK MIXING

1. In three separate ceramic vats, melt together the solvents, melt together the oils, and melt together the fats and waxy materials.
2. Add together the solvent solution and the oils. Mix thoroughly.
3. Add the pigments to the solvent/oil solution. This is the pigment mass.
4. Run the pigment mass mixture through a roller mill to remove any graininess from the mixture.
5. Place the mixture in the stirring machine for several hours.
6. Place the mixture in the commercial vacuum apparatus to remove all the air introduced in Step 5.
7. Add the pigment mass to the hot wax/fat solution.
8. Agitate the mixture thoroughly to remove air bubbles.
9. Strain the mixture thoroughly.

1. As a mixer in a cosmetics factory, you are in charge of melting and mixing the raw ingredients for lipstick. According to this procedures list, when should you use the roller mill?

   A. immediately after adding the pigments

   B. immediately before mixing together the solvent solution and the oils

   C. immediately after the mixture is stirred in the stirring machine

   D. immediately after the fats and the waxy materials are combined

   E. immediately after using the vacuum apparatus

2. According to the procedures list, what must be done just before the mixture is strained?

   A. The pigments must be added to the solvent/oil solution.

   B. The mixture must be stirred for several hours.

   C. The solvent solution and the oils must be mixed together.

   D. The mixture must be agitated to remove any air bubbles.

   E. The pigment mass mixture must be run through the roller mill to remove graininess.

Copyright © The McGraw-Hill Companies, Inc.

TO FREEZE DRY COFFEE, FOLLOW THESE STEPS:

1. Brew coffee by passing softened water through a series of columns of ground coffee beans. The first few columns are heated to 300°F and highly pressurized, and the remaining columns hold the water at boiling point (212°F) to enhance flavor.

2. The coffee is then passed through a heat exchanger to cool it down to 40°F. As a result of this cooling process, the coffee becomes 20% to 30% solid.

3. The coffee is further concentrated by processing it through a centrifuge, which separates the water from the coffee extract.

4. Next, the coffee extract is treated to remove as much oxygen as possible in order to preserve its aroma and flavor. This occurs via foaming gases like carbon dioxide and nitrogen through the extract before it is dehydrated.

5. Dehydration is achieved by first freezing the coffee extract to 20°F and then further cooling the resulting slush to −40°F in a very rapid process (one that takes approximately 90 seconds).

6. Now solid ice, the coffee extract is broken up and ground into particles.

7. The particles are transferred to a vacuum chamber. The mixture is dried out thoroughly. All the ice in the mixture is vaporized.

3. As a machine operator in a factory that makes instant coffee, you are asked to operate the centrifuge. You read the list of steps of the freeze-drying process. What should the consistency of the coffee be before it is processed in the centrifuge?

   A. solid ice

   B. liquid

   C. 20% to 30% solid

   D. gas

   E. watery

4. The next day, you are told to work on the machine that dehydrates the coffee extract. What temperatures must the coffee extract achieve during the two chilling processes in your machine?

   A. 300° F and 212° F

   B. 212° F and 40° F

   C. 40° F and −40° F

   D. 40° F and −20° F

   E. 20° F and −40° F

ANSWER KEY

Item 1: **A** immediately after adding the pigments

Item 2: **D** The mixture must be agitated to remove any air bubbles.

Item 3: **C** 20% to 30% solid

Item 4: **E** 20° F and −40° F

Copyright © The McGraw-Hill Companies, Inc.

SKILL

4

Read and
Understand
Information
in Workplace
Documents

**Follow
Instructions
from
Workplace
Documents**

Define and Use
Words in the
Workplace

Understand
and Follow
Policies and
Procedures
in Workplace
Documents

APPLY INSTRUCTIONS TO UNIQUE SITUATIONS

A set of instructions may call for different actions in different situations. For example, an environmental engineer working at a waterfront site must consider how being near water will change how waste is disposed of. The engineer must modify the steps taken to suit the unique situation.

CLOTHES DO NOT RINSE PROPERLY

There are many reasons why clothes do not rinse properly. If clothing is coming out of the spin cycle too wet, look for the following:

- Check water hose connections. If hoses are not secured to the machine properly, this could cause a discrepancy in water flow. This could result in too much water being used during the wash cycle, which is then unable to fully drain during the spin cycle.

- Check drain hose. If the drain hose is clogged or kinked, this could cause an interruption in the draining of wash cycle water and cause the wash water to remain in the wash drum during the spin cycle.

If these steps do not identify the source of the problem, more intrusive repair methods may be required. Do not attempt any on-site repairs without confirming the cause of the problem. Contact your supervisor or mentor before disassembling any appliance.

1. You are an appliance repairer helping a customer who is having problems with his washing machine's rinse cycle. The customer is complaining that his clothes are coming out of the washing machine too wet. You have checked the hose connections and determined that they are secure. What else should you check before proceeding further with any repairs?

   A. the spin cycle

   B. the water flow

   C. the machine's troubleshooting guide

   D. the drain hose

   E. the wash drum

2. You have checked both the hose connections and the drain hose for clogs. You run a test and find that clothing is still too wet after the rinse cycle and determine that you need to disassemble the machine to find the cause of the problem. What must you do before disassembling the machine?

   A. Contact your supervisor.

   B. Receive payment from the customer.

   C. Have the customer bring the machine to the repair shop.

   D. Reduce the amount of detergent used in each load.

   E. Secure the hose connections.

Copyright © The McGraw-Hill Companies, Inc.

BOTTLING AND LABELING VARIATIONS

Pacquette Cocktail Sauce is sold in all 50 United States as well as 24 other countries and areas. To make the product more attractive to consumers in different markets, we vary the appearance of the bottles, labels and caps according to prevailing preferences and tastes.

Before each run, check the destination of the order and confirm that these bottling/labeling conventions are being followed:

- For destinations outside the United States (except Africa), use the Paquette Cocktail Sauce label with English, Spanish, and French translations, the dark green glass bottles, and the green caps.

- For destinations within the 48 contiguous United States, use the Corky's Cocktail Sauce label with English only, the clear glass bottles, and the white caps.

- For destinations within Alaska, Hawaii, or Puerto Rico, use the Paquette Cocktail Sauce label with both Spanish and English translations, the clear glass bottles, and the green caps.

- If an order is destined for sale in Africa, use the Paquette Cocktail Sauce label with both French and English translations, the clear glass bottles, and the white caps.

3. You operate equipment at a factory that makes bottled food products. You are preparing the equipment to bottle an order that will be sold in the African nations of South Africa and Botswana. What label should you use for the order?

   A. Paquette label with English only

   B. Corky's label with English only

   C. Corky's label with English and French translations

   D. Paquette label with English and French translations

   E. Paquette label with English and Spanish translations

4. Next, you are preparing the equipment to bottle an order that will be sold in the southeastern region of the United States. What labels, bottles, and caps should you use?

   A. Paquette label with English only, clear glass bottles, green caps

   B. Corky's label with English only, clear glass bottles, white caps

   C. Corky's label with English and French, dark green glass bottles, white caps

   D. Corky's label with English and French, clear glass bottles, green caps

   E. Paquette label with English and French, dark green glass bottles, green caps

ANSWER KEY

Item 1: **D** the drain hose

Item 2: **A** Contact your supervisor.

Item 3: **D** Paquette label with English and French translations

Item 4: **B** Corky's label with English only, clear glass bottles, white caps

Copyright © The McGraw-Hill Companies, Inc.

DETERMINE THE MEANING OF NEW WORDS

Manufacturing workers occasionally come across words whose meaning is unclear. Some may be defined in the text, while others require the reader to discover the meaning. For example, a new worker at a paper factory must find the meanings of unfamiliar words in a procedures manual to be able to do the job. The context surrounding the word and the reader's background knowledge can help clarify the word's meaning.

According to Standard 29 CFR 1910.1(g), Federal employees working in establishments of private employers (such as meat and poultry plants) are covered by their agencies' occupational safety and health programs. Although an agency may not have the authority to require abatement of hazardous conditions in a private sector workplace, the agency head must ensure safe and healthful working conditions for his/her employees. This shall be accomplished using administrative controls, personal protective equipment, or withdrawal of Federal employees from the private-sector facility to the extent necessary to ensure the protection of the employees.

1. You are a federal employee who is working onsite as a quality control technician in a poultry processing plant. You have been given these guidelines to read. What is the meaning of the word **hazardous**?

    A. wet

    B. dangerous

    C. noisy

    D. frightening

    E. dirty

2. What is the meaning of the term **private sector**?

    A. run by the government

    B. large

    C. foreign owned

    D. not owned by the government

    E. dangerous

Copyright © The McGraw-Hill Companies, Inc. TEXT: U.S. Department of Agriculture, Food and Safety Inspection

MANUFACTURING PRODUCTION OCCUPATIONS

The jobs available in machinery manufacturing are varied and constantly changing along with the technology used in this field. Most workers in this field are required to have a high-school diploma or the equivalent. Workers may also receive some on-the-job training. Skilled production workers, such as engineers and machinists, usually need to have previous on-the-job experience or have been trained as part of a program at a community or technical college. Apprenticeship programs that train workers new to the field are offered by some employers. These apprenticeships can last between one and five years, depending on the area of specialty. An apprenticeship offers on-the-job training and classroom instruction, either with the employer or at a technical school. Apprenticeship areas include tool designing, programming of computer-controlled machines, blueprint reading, mathematics, and electronics. In addition to learning technical skills, apprentices also learn about quality control, safety, and communication.

After gaining experience, a worker may have the opportunity to move into higher-level positions. Advancement is based on experience and merit, so with hard work and constant improvement, entry-level workers can look forward to advancing to significantly higher-skilled jobs.

3. You are considering taking a job as a tool and die maker. Before doing so, you read the *Career Guide to Machinery Manufacturing,* which includes this passage. What do you think **apprenticeship programs** do?

   A. Screen applicants for jobs.

   B. Combine on-the-job training with classroom instruction.

   C. Send job applicants to college.

   D. Help job applicants get a high school diploma.

   E. Send job applicants listings of job openings.

4. After reading the passage, what do you think **merit** means in this context?

   A. a type of diploma

   B. job experience

   C. good job performance

   D. a college degree

   E. participation in an apprenticeship program

ANSWER KEY

Item 1:   **B** dangerous

Item 2:   **D** not owned by the government

Item 3:   **B** Combine on-the-job training with classroom instruction.

Item 4:   **C** good job performance

Copyright © The McGraw-Hill Companies, Inc.

UNDERSTAND UNIQUE WORDS AND ACRONYMS

Acronyms (words made from the initials of several words) and abbreviations may sometimes be used without explanation in work situations. For example, an environmental compliance inspector reading federal rules about environmental impact might come across the acronym LDR (Land Disposal Restriction). To understand such terms, readers should use prior knowledge or study the surrounding text to determine their meaning.

TO ALL EMPLOYEES:

An important "Guidance" document has recently been issued by the Food and Drug Administration and its subagency, the Center for Drug Evaluation and Research (CDER) regarding the risk of Melamine contamination in our industry. Safety engineers and product safety managers **must** be aware of the following main points of the guidance:

- All materials and components must be tested before they are released for use, according to the current good manufacturing practice (CGMP) regulations. Manufacturers must identify any component they intend to use to manufacture or prepare a drug product as at-risk for melamine contamination.
- Liquid chromatography triple quadruple tandem mass spectrometry (LC-MS/MS) and gas chromatography/mass spectrometry (GC-MS) should be used in measuring melamine contamination in foods, per the FDA.

---

1. You are the product safety manager at a plant that manufactures aspirin. You receive this notice about the dangers of the contaminant melamine. Which of the following acronyms stands for the subagency of the Food and Drug Administration that released this notice?

   A. FDA

   B. CDER

   C. CGMP

   D. PPM

   E. QC

2. What is the main requirement of CGMP regulations?

   A. All components be tested before they are released.

   B. All components must have melamine in them.

   C. All distributors must do their own quality control.

   D. The FDA must test all pharmaceuticals for melamine.

   E. The QC process must be repeated at least twice.

Copyright © The McGraw-Hill Companies, Inc.

MEMO

From: Bill Akin-Mellis

To: Process Engineers

Please read the following two paragraphs from the document and evaluate whether these procedures and changes would be useful in our plant.

Install air saver nozzles on press machine blowoff lines — Several of the presses use a continuous stream of compressed air blown through two $\frac{3}{8}$-inch open pipes to detach parts from the dies. The assessment team recommended installing high-thrust nozzles to reduce compressed air usage. Air-saver nozzles work by entraining ambient air into the flow of compressed air. . . .

Install a controlled cooling system for parts whose heat treating is currently outsourced—The assessment team recommended options for in-house controlled cooling of forged parts that are currently being outsourced for heat treating. The options were to use a batch-type cooling system (1) in which parts are placed in bins and cooled under controlled temperature immediately after being forged, and (2) to control the cooling of parts produced from the ring rolling machines, and two continuous (spiral) systems to handle single parts produced directly from the Hatebur presses. If cooling bins are used, the batch-type systems should feature high-convection recirculating air flow to ensure uniform cooling of all the parts.

3. As an industrial process engineer at a steel plant, you evaluate procedures. This morning you received a memo from your boss. What does the word **dies** mean as it is used in the second paragraph of the memo?

   A. stops living

   B. devices for cutting or stamping metal

   C. tints or pigments used for changing the color materials

   D. parts that have stopped functioning

   E. high-thrust nozzles

4. What does the word **uniform** mean in the last paragraph of this memo?

   A. clothing

   B. even

   C. soft

   D. hard

   E. different

ANSWER KEY

Item 1: **B** CDER

Item 2: **A** All components be tested before they are released.

Item 3: **B** devices for cutting or stamping metal

Item 4: **B** even

Copyright © The McGraw-Hill Companies, Inc. TEXT: U.S. Department of Energy, Energy Efficiency and Renewable Energy

UNDERSTAND AND APPLY TECHNICAL TERMS AND JARGON

Some workplace documents use technical terms and jargon that are specific to an occupation. Food safety managers may encounter technical terms, such as the names of bacteria, in a government food code document. They must be able to interpret the meanings of these terms and apply them to the situation at hand.

MANUFACTURING SPECIALIST, TROPHYWARE

Job Description:

Responsible for setting up and operating a variety of automatic or semiautomatic machines used in the production of the company's products (silverplated trophyware) including activities such as masking and etching products. The etching work requires great attention to detail, as you are cutting into the silver—and mistakes cannot be corrected. Makes precise measurements so cuts are accurate to tolerances of a few thousandths of an inch. Must be available to work any shift.

1. Some work on holidays and weekends is required.

2. Night shift employees receive a premium. (15% 7 p.m.–7 a.m.)

3. If you wear glasses or contacts, we will provide you with a prescription pair of glasses once you bring in a prescription less than two years old.

4. Expect to be on your feet 100% of the time.

Requirements:

Excellent visual acuity, manual dexterity, handwriting, and drawing skills. Extreme attention to detail.

Location: Youngstown, OH

Type: Full-time

1. As an experienced metalworker with considerable artistic talent, you are considering applying for this job. What does **etching** mean?

   A. engraving

   B. driving

   C. stitching

   D. rolling

   E. bouncing

2. In the context of this job description, what does the word **tolerances** refer to?

   A. difficult assignments

   B. lengths

   C. maximum size of errors

   D. minimum size of errors

   E. number of cuts

Copyright © The McGraw-Hill Companies, Inc.

MEMORANDUM

To: Press Operators, First and Second Shift

From: Customer Service Manager

Re: Customer Press Check, Monday, 2/14

Press operators, please be aware that a customer will be in the plant this Monday, 2/14, to press check her magazine's April issue. Also, she will be using contact proofs to check the color reproduction of the advertising pages of the issue, and not the editorial pages. This customer likes to bring along her own mockup of the issue to make sure that all the pages are in the correct sequence, so please be patient as she reviews the make-ready sheets.

3. You are a press operator at a printing plant that specializes in magazines. You have received a memorandum making you aware of a customer visit this Monday. After reading the memo, what do you think is the purpose of **contact proofs**?

   A. to verify color accuracy

   B. to check for typographical errors

   C. to verify the magazine size

   D. to verify the type of paper

   E. to help draft the business contract

4. What does the word **mockup** mean in this memo?

   A. fake image

   B. ideal image

   C. rough prototype

   D. complementary review

   E. list of problems

ANSWER KEY

Item 1:   **A** engraving

Item 2:   **C** maximum size of errors

Item 3:   **A** to verify color accuracy

Item 4:   **C** rough prototype

Copyright © The McGraw-Hill Companies, Inc.

Read and
Understand
Information
in Workplace
Documents

Follow
Instructions
from Workplace
Documents

Define and Use
Words in the
Workplace

**Understand
and Follow
Policies and
Procedures
in Workplace
Documents**

APPLY WORKPLACE POLICIES AND PROCEDURES

Many manufacturing workers receive a policies and procedures manual when they begin work. It is important not just to understand the text of the manual, but to also apply the policies to their actual work situation. For example, understanding the reasons for a factory's workplace safety principles can help workers deal with potentially dangerous events.

MEMO: INCLEMENT WEATHER POLICY

All employees are expected to make every effort to report to work on time despite weather problems. Inclement weather conditions require employees to make decisions regarding their safety when traveling to and from the workplace. Employees who do not feel it is safe to travel should contact their supervisor using approved procedure. Employees who do not report to work during periods of inclement weather may use accumulated personal or vacation time for their absence. If the employee does not have any personal or vacation time left, then the absence is counted as leave without pay. Employees who make the effort to leave home on time and report to work within a reasonable period should not be required to take leave for their tardiness.

1. You are a supervisor in a printing plant that remains open even during periods of bad weather. This morning you received a memo from Human Resources regarding weather-related absences and tardiness. What is the principle behind this policy?

   A. Weather permitting, employees are expected to be at work.

   B. Missing a shift is cause for termination.

   C. Employees are not expected to come to work during storms.

   D. No employee should miss work because of the weather.

   E. Lateness because of bad weather is not acceptable.

2. You arrived 30 minutes late one morning because of flooding, even though you left home on time. You mark 30 minutes as personal time on your time sheet, but your supervisor deletes it from your time sheet. Why did your supervisor do this?

   A. Your supervisor feels sorry for you.

   B. Your supervisor does not accept your reason for being late.

   C. Tardiness on a day with bad weather may be excused.

   D. You have not accumulated any leave time.

   E. The time will be charged as leave without pay.

Copyright © The McGraw-Hill Companies, Inc.

PRODUCTION PROCESSES: START-UP

A sand and gravel, or crushed stone, plant consists of a number of interdependent production processes. Therefore, it is important to know how the plant operates in order to prevent a massive pileup of material at a transfer point during plant start-up. An improper start-up sequence can damage plant equipment, and also increase the risk of injury. Freshwater, sand, and other pumps are started first. Other plant equipment must be started in reverse order of material flow, beginning with the finished product conveyor, and working back through the primary hopper feeder belt.

In addition to following the proper start-up sequence, the person starting the plant must take every precaution to ensure that other people are clear of equipment before the equipment is started.

3. As a new supervisor at a sand and gravel plant, you are reviewing this document to help you understand the start-up sequence of plant equipment. You are now ready to start your shift and need to consider the start-up sequence. Which of the following is true?

   A. If you start the pumps first, you could damage the plant equipment.

   B. You can start all of the plant equipment at the same time— the sequence does not matter.

   C. If you start the primary hopper feeder belt first, you could damage the plant equipment.

   D. If you start the primary hopper feeder belt last, you could damage the plant equipment.

   E. The pumps should be started last, after all the other equipment has been started.

4. The next day, you arrive at the plant and the start-up process has already begun. Suddenly, a pileup occurs on one of the conveyors. According to the document, which of the following events could have led to the pileup?

   A. The freshwater, sand, and other pumps were started first, before all the other equipment.

   B. The primary hopper feeder belt was started before the finished product conveyor.

   C. The finished product conveyor was started before the primary hopper feeder belt.

   D. Plant equipment was started in reverse order.

   E. Workers were clear of the equipment before the start-up.

ANSWER KEY

Item 1: **A** Weather permitting, employees are expected to be at work.

Item 2: **C** Tardiness on a day with bad weather may be excused.

Item 3: **C** If you start the primary hopper feeder belt first, you could damage the plant equipment.

Item 4: **B** The primary hopper feeder belt was started before the finished product conveyor.

Copyright © The McGraw-Hill Companies, Inc. TEXT: U.S. Department of Labor, Mine Safety and Health Administration

UNDERSTAND THE RATIONALE BEHIND WORKPLACE POLICIES

As with any industry, workplace policies in the manufacturing industry are created for a reason. A quality systems control manager in an airplane factory must be able to explain why parts must exactly meet the measurements called for. This knowledge helps workers to ensure the policies are being followed in the proper manner.

MEMORANDUM

To: First Shift Assemblers   From: Supervisor

It has come to our attention that a surprising number of the handbags we have produced over the past three months have had too many unnecessary stitches. As you know, the stitches permanently damage the leather, and even if the threads are removed, the holes created by the needles will not close. The quality control department has noted that the number of defective bags with excessive stitching has exceeded 5% of all bags produced during that period.

In order to correct this issue, we will hold a one-hour training session on proper stitching procedures next Monday morning at 8 a.m., one hour before your normal shift start time. (You will be compensated for the extra hour of work.) Please report to the break room at that time.

1. You are an assembler in a handbag factory and have receive this memo. What is the rationale behind the training session that you have been told to attend?

   A. Assemblers will damage fewer bags if they learn proper stitching techniques.

   B. Assemblers show up late too often and need retraining.

   C. Assemblers need to spend less time in the break room.

   D. Assemblers should be compensated for extra hours.

   E. Assemblers need training on how to remove excessive stitches.

2. What is most likely the rationale for having the training session at 8 a.m.?

   A. The session will wake workers up so they will be more alert when work begins.

   B. It is just after breakfast, so the assemblers will not be distracted because they are hungry.

   C. It is before the shift begins, so it will not interfere with the assemblers' work.

   D. It is early in the morning, so the assemblers will be relaxed.

   E. It is before the shift begins, so the company will not have to pay the assemblers for their time.

Copyright © The McGraw-Hill Companies, Inc.

EXPLOSIVES SAFETY AND SECURITY GUIDELINES

- Regular inventories should be conducted several times a year to ensure that no thefts or loss of explosive materials has occurred.

- Keep federal, state, and local agency telephone numbers accessible in order to report a theft, loss, or suspicious activity.

- Make sure your storage site is secure.

- Allow only essential personnel to have access to explosives storage sites. This limits the possibility of magazines being stolen or tampered with.

- Fences, floodlights, alarms, security cameras, locked gates, or other security devices at the site can help you better monitor the location. In highly visible places, post security signs to deter unauthorized access.

- Keep in regular contact with local law enforcement and get to know the officers who routinely patrol your area. Tell them about your business hours so that they can alert you when people are present during hours when the site is closed.

- Make sure your employees are well-trained in securing your explosives. They need to recognize suspicious customers or unusual activities, and to be able to respond to thefts, losses, or emergencies.

3. As head of security at an explosives factory, you need to follow best practices to ensure safety. What is the rationale behind this document?

   A. Explosives are expensive and tempting targets for thieves.

   B. Explosives must not fall into the wrong hands, because they can be used to cause great harm.

   C. Explosives must be tightly controlled by the government.

   D. Explosives are vulnerable to theft by vendors and customers, so those parties must be watched.

   E. Explosives should only be handled by essential personnel.

4. Why do the best practices suggest that regular inventories should be performed?

   A. to make sure the explosives still work

   B. to make sure employees stay busy

   C. to make sure none of the explosives have been stolen

   D. to make sure the explosives meet regulations

   E. to make sure only vendors and employees have access to the explosives

ANSWER KEY

Item 1: **A** Assemblers will damage fewer bags if they learn proper stitching techniques.

Item 2: **C** It is before the shift begins, so it will not interfere with the assemblers' work.

Item 3: **B** Explosives must not fall into the wrong hands, because they can be used to cause great harm.

Item 4: **C** to make sure none of the explosives have been stolen

Copyright © The McGraw-Hill Companies, Inc.

LOCATING INFORMATION

To succeed in a manufacturing career, you must be able to effectively locate information. Information comes in a variety of forms, including tables, graphs, maps, and diagrams. You may need to locate this information in graphics on a computer screen, in a document, or even posted on a bulletin board or wall.

Locating information means more than just finding it. It also means understanding it and making use of it in the job you do each day. It may also mean finding missing information and adding it to a graphic.

On the following pages, you will encounter a variety of workplace graphics. You will be asked to find important information in these graphics. In some cases you must interpret information in these graphics. For example, you may need to compare data, summarize it, or sort through distracting information.

When you read a question on the following pages, think about what is being asked and how you might find the answer. Look carefully at the graphic, focusing on the information you are asked to find or the steps you are asked to take. After you have chosen an answer, look back to make sure you have answered the question being asked.

Learning these key locating information skills will speed your path to advancement in the manufacturing industry.

Copyright © The McGraw-Hill Companies, Inc. PHOTO: Fuse/Getty Images.

KEY SKILLS FOR CAREER SUCCESS

Here are the topics and skills covered in this section and some examples of how you might use them to locate information in different types of graphics.

TOPIC	SKILL
Locate and Compare Information in Graphics	1. Find Information in Workplace Graphics 2. Enter Information into Workplace Graphics

Example: As a shipping and receiving clerk, you may need to review or fill out shipping orders for shipments that are sent and received.

Analyze Trends in Workplace Graphics	3. Identify Trends in Workplace Graphics 4. Compare Trends in Workplace Graphics

Example: As a safety technician, you may need to identify trends shown in graphs in a report on industry safety.

Use Information from Workplace Graphics	5. Summarize Information in Workplace Graphics 6. Make Decisions Based on Workplace Graphics

Example: As a plant manager, you may need to compare charts that describe the offerings of several vendors and select the company that can provide materials needed at the lowest cost.

Copyright © The McGraw-Hill Companies, Inc.

FIND INFORMATION IN WORKPLACE GRAPHICS

When reading a workplace graphic, manufacturing workers must know what information to look for. For example, a machinist must understand a diagram to make the particular part he or she is about to produce. Workers must be able to sift through irrelevant or distracting information to find what is needed.

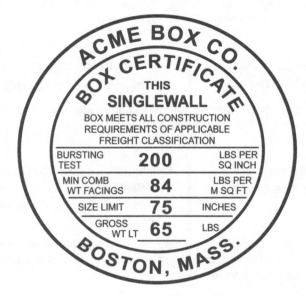

1. You are in charge of packaging for a textile plant. You are selecting a box for 80-pound shipments of queen-size sheet sets to distributors nationwide. Each sheet set package is 4 inches thick, 10 inches long, and 7 inches wide. Why is the box described by the box certificate inappropriate for the job?

   A. The box is too narrow to hold all the sheet sets.

   B. The sheet sets are too heavy for the box.

   C. There are too many sheet sets to fit in the box.

   D. The sheet sets are too long for the box.

   E. The sheet sets are too thick for the box.

2. Which of the following items could be shipped in the box described in the box certificate?

   A. a 60-pound queen-size sheet set shipment

   B. a 70-pound queen-size sheet set shipment

   C. a 70-pound twin-size sheet set shipment

   D. an 84-pound twin-size sheet set shipment

   E. a 90-pound queen-size sheet set shipment

Copyright © The McGraw-Hill Companies, Inc.

Using a Toolmaker's Vise

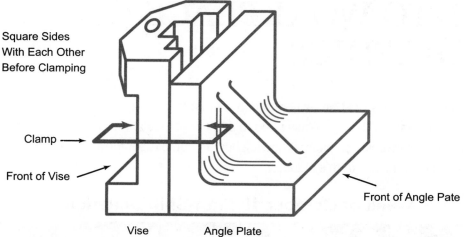

Square Sides
With Each Other
Before Clamping

Clamp →

Front of Vise →

Front of Angle Pate

Vise   Angle Plate

3. You are a tool and die worker using a small precision toolmaker's vise to hold in place a small piece of machinery as you connect it to a larger piece of machinery. The graphic illustrates how the vise's body is pressed against an angle plate. According to the graphic, where should you place the clamp to hold the vise shut?

A. along the sides of the vise and angle plate

B. along the front of the angle plate

C. along the front of the vise

D. on top of the vise and the angle plate

E. below the vise and the angle plate

4. According to the graphic, what must you check before placing the clamp?

A. The height of the angle plate is the same as the vise.

B. The width of the angle plate is the same as the vise.

C. The sides of the angle plate and the vise are squared with each other.

D. The angle plate is wider than the vise.

E. The angle plate is narrower than the vise.

ANSWER KEY

Item 1:   **B** The sheet sets are too heavy for the box.
Item 2:   **A** a 60-pound queen-size sheet set shipment
Item 3:   **A** along the sides of the vise and angle plate
Item 4:   **C** The sides of the angle plate and the vise are squared with each other.

Copyright © The McGraw-Hill Companies, Inc.

SKILL

2

Locate and
Compare
Information
in Graphics

Analyze Trends
in Workplace
Graphics

Use Information
from Workplace
Graphics

ENTER INFORMATION INTO WORKPLACE GRAPHICS

It may be necessary at times to add information to a workplace graphic. As a production supervisor, you may need to add the names and shift hours of floor workers to a weekly calendar. Knowing how to complete informational graphics is an important skill in this industry.

Week of October 10 First-Shift Schedule

	Monday	Tuesday	Wednesday	Thursday	Friday	Saturday	Sunday
Deshawn	On-duty	OFF	On-duty	On-duty	OFF	On-duty	On-duty
Fabrice			OFF				
Fred	On-duty	OFF	On-duty	On-duty	On-duty	OFF	On-duty
Janelle	On-duty	On-duty	On-duty	OFF	On-duty	On-duty	OFF
Lita	OFF	On-duty	On-duty	On-duty	OFF	On-duty	On-duty
Mauricio	On-duty	On-duty	OFF	On-duty	On-duty	OFF	On-duty
Quinton	On-duty	OFF	On-duty	On-duty	On-duty	OFF	On-duty
Robert	OFF	On-duty	On-duty	OFF	On-duty	On-duty	On-duty
Rosa	OFF	On-duty	On-duty	On-duty	OFF	On-duty	On-duty
Sylvia	On-duty	On-duty	OFF	On-duty	On-duty	On-duty	OFF

1. You are a department supervisor in a factory that makes turbines. You supervise the first shift, which includes 10 employees total. Each employee gets two days off a week. At least 7 employees must be working each day. You are almost finished with the weekly schedule, but you still have to choose another day off for Fabrice. Which days would be possible days off for Fabrice?

   A. Friday or Saturday

   B. Monday or Tuesday

   C. Monday or Friday

   D. Thursday or Sunday

   E. Tuesday or Thursday

2. Rosa decides that she wants to work on Monday and take Wednesday off instead. Which employees could she switch days off with?

   A. Fabrice, Mauricio, or Sylvia

   B. Janelle, Robert, or Lita

   C. Lita, Quinton, or Rosa

   D. Robert, Deshawn, or Janelle

   E. Deshawn, Quinton, or Janelle

Copyright © The McGraw-Hill Companies, Inc.

Pasteurization Process

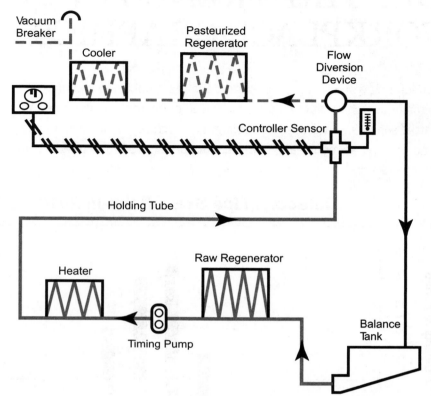

3. You operate a pasteurization machine at a factory that makes yogurt. The diagram shows how raw milk flows through your machine. At what point does the milk pass through the heater?

A. immediately after passing through the balance tank

B. immediately before entering the balance tank

C. immediately before passing through the timing pump

D. immediately after passing through the timing pump

E. immediately after passing through the holding tube

4. Where does the milk go after it leaves the balance tank?

A. to the raw regenerator

B. to the flow diversion device

C. to the controller sensor

D. to the cooler

E. to the timing pump

ANSWER KEY

Item 1:   **D** Thursday or Sunday

Item 2:   **A** Fabrice, Mauricio, or Sylvia

Item 3:   **D** immediately after passing through the timing pump

Item 4:   **A** to the raw regenerator

Copyright © The McGraw-Hill Companies, Inc.

Locate and
Compare
Information
in Graphics

**Analyze
Trends in
Workplace
Graphics**

Use Information
from Workplace
Graphics

IDENTIFY TRENDS IN WORKPLACE GRAPHICS

Manufacturing workers must sometimes analyze graphics to identify trends. They might search for data that has increased or decreased over time. An industrial safety and health engineer might analyze tables of accident statistics in a factory over several years. Being able to identify common trends from several pieces of data can help with a variety of jobs in this industry.

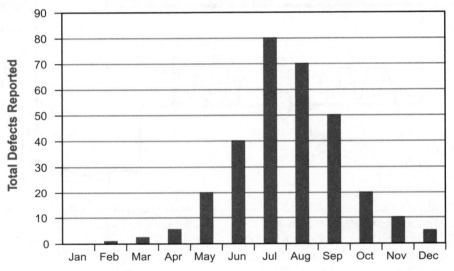

Defects in Ice Skate Pairs in 2010

1. As a shift manager in a production plant that makes ice skates, you have received the defect report for your shift's production over the past year. In what month were the defects at their highest point?

   A. January

   B. March

   C. April

   D. June

   E. July

2. By how much did the defect rate change between May and June?

   A. It doubled.

   B. It was reduced by half.

   C. It stayed the same.

   D. It fell slightly.

   E. There were no defects in June.

Copyright © The McGraw-Hill Companies, Inc.

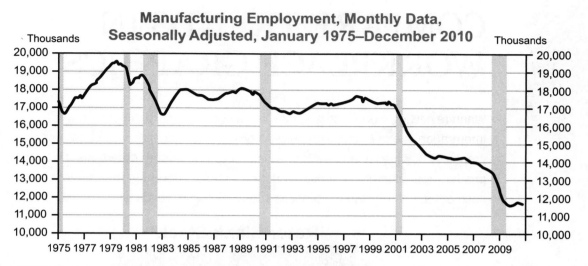

Manufacturing Employment, Monthly Data, Seasonally Adjusted, January 1975–December 2010

NOTE: Shaded areas represent recessions as determined by the National Bureau of Economic Research (NBER).

3. You are an experienced machine operator considering a new job in manufacturing. What trend do you see within manufacturing jobs?

   A. The number of jobs has been steadily increasing since 1991.

   B. The number of jobs has fluctuated greatly since 1989.

   C. The number of jobs has remained steady since 1975.

   D. The number of manufacturing jobs has started decreasing in recent years.

   E. The number of manufacturing jobs has been steadily decreasing since 2001.

4. How would you describe the trend for manufacturing jobs between 1975 and 1983?

   A. Jobs decreased sharply until 1978, when they began increasing again.

   B. Jobs remained unchanged between 1975 and 1983.

   C. Jobs increased each year from 1975 to 1983.

   D. Jobs rose steadily until 1979, and then decreased.

   E. Jobs decreased steadily each year from 1975 to 1983.

ANSWER KEY

Item 1: **E** July

Item 2: **A** It doubled.

Item 3: **E** The number of manufacturing jobs has been steadily decreasing since 2001.

Item 4: **D** Jobs rose steadily until 1979, and then decreased.

Copyright © The McGraw-Hill Companies, Inc.

COMPARE TRENDS IN WORKPLACE GRAPHICS

When reviewing workplace graphics, it may be necessary to compare information in one or more graphics. A production manager might compare graphics showing various ways of assembling a product to determine the best process. Manufacturing workers must know how different graphics relate to each other, and be able to compare information and trends within them.

Rotor Blank Stock

Rotor Type	Blank Castings in Stock
EJ2165-a	2691
L4310	8340
Ch338A24	5165
To53JK22	3020
M2592-ar3	1745

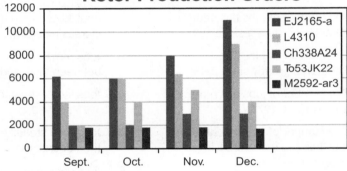

Rotor Production Orders

1. As a production manager, you make sure you have enough blank castings in stock to fill the rotor orders each month. It is late December and you are comparing recent production of rotors to your current castings in stock to prepare your next order. Based on the charts, which kinds of castings can you be reasonably certain you will have to reorder before the end of January?

   A. EJ2165-a

   B. L4310

   C. Ch338A24

   D. To53JK22

   E. M2592-ar3

2. Based on the production trends and your current stock, which type of castings are you reasonably certain you won't need to reorder in January?

   A. EJ2165-a

   B. L4310

   C. Ch338A24

   D. To53JK22

   E. M2592-ar3

Copyright © The McGraw-Hill Companies, Inc.

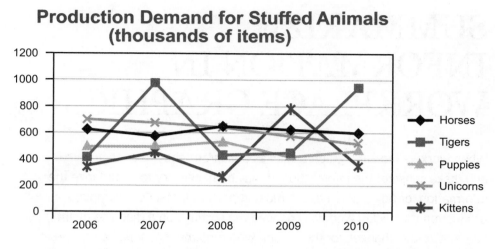

Production Demand for Stuffed Animals (thousands of items)

Legend:
- Horses
- Tigers
- Puppies
- Unicorns
- Kittens

3. As the production manager of a stuffed toy company, you are studying patterns in your company's manufacturing production over a period of five years for five different types of stuffed animals—horses, tigers, puppies, unicorns, and kittens. Demand for these animals has fluctuated over the past few years. While the tigers production demand reached its second-highest peak in 2010, another animal hit its second-lowest demand for production in the same year. Which is it?

A. horses

B. tigers

C. puppies

D. kittens

E. unicorns

4. When hit movies feature an animal, your company often sees a spike in manufacturing demand for that type of animal. What types of animals most likely starred in hit movies between 2006 and 2010?

A. horses

B. horses and tigers

C. puppies

D. unicorns and puppies

E. kittens and tigers

ANSWER KEY

Item 1: **A** EJ2165-a

Item 2: **C** Ch338A24

Item 3: **D** kittens

Item 4: **E** kittens and tigers

Copyright © The McGraw-Hill Companies, Inc.

SKILL 5

Locate and Compare Information in Graphics

Analyze Trends in Workplace Graphics

Use Information from Workplace Graphics

SUMMARIZE INFORMATION IN WORKPLACE GRAPHICS

When workers look at a graphic such as a diagram, they need to analyze and make sense of the information. It may be necessary to summarize the information, or boil it down to the most important facts. For example, a production technician who has changed the settings on a machine might summarize the changes for the person who runs the machine. Being able to summarize allows workers to make sense of varying information.

Oregon Labor Force by Area

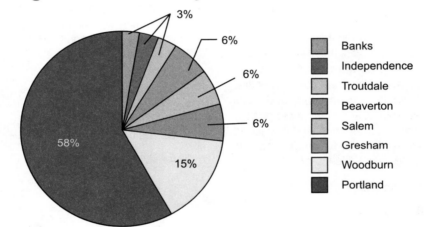

Legend:
- Banks
- Independence
- Troutdale
- Beaverton
- Salem
- Gresham
- Woodburn
- Portland

1. You are a shift manager at a paper mill in Oregon. You are trying to recruit qualified workers for your shift and want to draw from the local population. Looking at the chart, what can be said in general about the best places in the state to find workers?

   A. They are available equally all over the state.

   B. Most of them come from Salem

   C. Banks, Independence, and Troutdale have the most laborers.

   D. Portland accounts for over half of the labor force.

   E. The chart does not indicate where the most workers come from.

2. What can be said about the labor force available in Beaverton, Salem, and Gresham?

   A. There are more workers available in Gresham.

   B. There is an equal number of workers in each of the three cities.

   C. There are twice as many available workers in Beaverton than in Salem or Gresham.

   D. Salem has 10% more available workers than Gresham and Beaverton.

   E. The available workers in the three cities have increased over the last 2 years.

Copyright © The McGraw-Hill Companies, Inc.

Shipping Costs Comparison: Seattle vs. Chicago

To/From	Seattle		Chicago	
	Cost	Days	Cost	Days
Atlanta	$14.35	4	$10.56	2
Denver	$11.20	3	$11.20	2
Houston	$12.85	3	$10.56	3
Indianapolis	$12.60	4	$9.50	2
Memphis	$12.60	3	$9.50	1
Miami	$14.35	4	$11.20	3
Portland	$8.15	1	$12.60	3
San Francisco	$10.56	2	$12.60	4
Average cost per package	**$12.69**		**$10.88**	
Average shipping days per package		3.3		2.1
Monthly cost 500 packages per day	**$139,601.00**		**$119,724.00**	
Monthly savings if shipping from Chicago	**$19,877.00**			
Annual savings if shipping from Chicago	**$238,524.00**			
(Costs are based on current shipping rates for a 20 pound package)				

3. You are a logistics manager for a shoe manufacturer whose warehouse is located in Seattle. Business is growing and orders are coming increasingly from other parts of the United States. You are considering relocating your warehouse to Chicago. What is the advantage, if any, of shipping packages from Chicago.

   A. There is no advantage.

   B. The average shipping cost per package will be lower, at $10.88.

   C. The average shipping cost per package will be lower, at $12.39.

   D. The annual savings will be $19,877.00.

   E. The annual savings will be $119,724.00.

4. How would a Chicago distribution center impact shipping times?

   A. It would have no impact.

   B. It would decrease shipping times by 2.4 days for each package.

   C. It would decrease shipping times by 3.3 days for each package.

   D. It would decrease the average shipping time to 1.2 days per package.

   E. It would reduce the average shipping time to 3.3 days per package.

ANSWER KEY

Item 1: **D** Portland accounts for over half of the laborers.

Item 2: **B** There is an equal number of workers in each of the three cities.

Item 3: **B** The average shipping cost per package will be lower, at $10.88.

Item 4: **D** It would reduce the average shipping time to 1.2 days per package.

Copyright © The McGraw-Hill Companies, Inc.

SKILL

6

Locate and
Compare
Information
in Graphics

Analyze Trends
in Workplace
Graphics

**Use
Information
from
Workplace
Graphics**

MAKE DECISIONS BASED ON WORKPLACE GRAPHICS

After analyzing the information in a workplace graphic, the next step is often to make a decision or take action as a result of the analysis. In the manufacturing industry, a quality control systems manager might add a step in how cell phones are tested if a flow chart shows that a problem is not being caught. Making the right decisions based on graphical information can help improve a business and make workers more effective in their jobs.

Error Code Troubleshooting - Vending Machines

Error Code	Potential Reasons	Potential Fixes (follow each step in order)
Out of Service	The door switch was left open longer than 30 minutes.	1. Check plug. 2. Door left open too long or product loaded into machine was too warm—leave closed another 15 minutes. 3. Power outage—check plug. 4. Install new door switch.
Button Jam	One of the keypad's selection buttons has been depressed for longer than 2 minutes.	1. Check plug. 2. Thoroughly clean all keypad selection buttons. 3. Replace keypad.
Product Sensor Block	The machine's product sensor, which indicates whether a product is left in the machine, is blocked.	1. Check plug. 2. Remove products that may have settled at the bottom of the machine. 3. Check and clear all sensor openings in the delivery bin. 4. Check the sensor harness connections.
Cleaning Needed	A reminder to have a thorough cleaning performed every six months.	If the last cleaning was less than six months ago: 1. Check plug. 2. Check delivery bin and bottom of machine for spillage. 3. Clean the machine thoroughly.

1. Your job at a vending machine factory is to test each machine to make sure it is functioning correctly. During a test, you get the error message "Button Jam." You check the plug and clean the keypad. The "Button Jam" message is still illuminated. What should you do next?

   A. Clean the screens.

   B. Replace the plug.

   C. Replace the keypad.

   D. Check the sensor harness connections.

   E. Replace the door switch.

2. You are testing a machine and get the error message "Out of Service." You unplug the machine and plug it in again. The "Out of Service" message is still illuminated. What should you do next?

   A. Clean the screens.

   B. Replace the plug.

   C. Replace the keypad.

   D. Wait 15 minutes.

   E. Replace the door switch.

Copyright © The McGraw-Hill Companies, Inc.

Order Form #1		
Quantity 500	**Date Order Placed** 2/24	**Date Order Needed** 3/24
Brace Arrangement:	Classic ~~Dreadnought~~	Jumbo
Frets:	12 14 ~~19~~	24
Strings:	Nylon ~~Steel~~	
Neck Wood:	Brazilian Indian Rosewood ~~Mahogany~~ Maple Rosewood	
Back and Sides Wood:	Mahogany ~~Koa~~	
Top Wood:	Red Spruce ~~Maple~~	Mahogany

Order Form #2		
Quantity 200	**Date Order Placed** 3/29	**Date Order Needed** 4/22
Brace Arrangement:	~~Classic~~	Dreadnought Jumbo
Frets:	12 14	19 24
Strings:	~~Nylon~~	Steel
Neck Wood:	Brazilian Indian Rosewood ~~Mahogany~~ Maple Rosewood	
Back and Sides Wood:	~~Mahogany~~ Koa	
Top Wood:	Red Spruce Maple	~~Mahogany~~

Dreadnought Guitar Schematics

Exploded View

3. As a scheduler for a guitar manufacturer, you notify the workers who build different pieces of guitars about orders. The drawing shows the location of top braces on an acoustic guitar. In order form #1, which worker needs to know what top brace arrangement was chosen?

   A. the Classic back builder

   B. the Classic sides builder

   C. the Dreadnought neck builder

   D. the Dreadnought top builder

   E. the Jumbo sides builder

4. The second order form is for 200 Classic acoustic guitars. The customer forgot to make a selection for one critical piece of information. Which craftsman cannot start work until you get that information?

   A. the back builder

   B. the sides builder

   C. the neck builder

   D. the top builder

   E. the tuning machine builder

ANSWER KEY

Item 1: **C** Replace the keypad.

Item 2: **D** Wait 15 minutes.

Item 3: **D** the Dreadnought top builder

Item 4: **C** the neck builder

Copyright © The McGraw-Hill Companies, Inc.

APPLIED MATHEMATICS

Using applied mathematics will help you succeed in the manufacturing career cluster. Some careers in this cluster, such as manufacturing production technician, require extensive mathematical skill. Others, such as solderers and brazers, require less complex mathematical skills. No matter what the career, math will be used in one way or another. Some key skills include multiplying and dividing, finding percentages, and adding fractions.

On the following pages, you will encounter a variety of applied math problems. Each item describes a real-life situation in a manufacturing career. You will be asked to solve the problems by applying your mathematical skills. You may need to use arithmetic, geometry, or measurement skills, for example.

When you read a question on the following pages, think about what is being asked and how you might find the answer. Read the problem carefully, focusing on the information you are asked to find or the steps you are asked to take. After you have chosen an answer, look back to make sure you have answered the question being asked.

By learning and practicing these key mathematical skills, you will put yourself in a better position to succeed in the manufacturing industry.

Copyright © The McGraw-Hill Companies, Inc. PHOTO: DEX IMAGE/Getty Images.

KEY SKILLS FOR CAREER SUCCESS

Here are the topics and skills covered in this section and some examples how you might use them to solve workplace problems.

TOPIC	SKILL
Perform Basic Arithmetic Calculations to Solve Workplace Problems	1. Solve Problems with Whole Numbers and Negative Numbers 2. Use Fractions, Decimals, and Percents to Solve Workplace Problems

Example: As a medical equipment repairer, you may need to use basic arithmetic when testing and measuring components and equipment.

Apply Computations to Solve Workplace Problems	3. Use General Problem Solving 4. Solve Problems in Geometry

Example: As a machine assembler, you may need to calculate areas and perimeters when selecting parts for assembly.

Solve Measurement Problems	5. Calculate with Conversions and Formulas 6. Manipulate Formulas to Solve Problems

Example: As an industrial engineering technician, you may need to use formulas when calculating a production plant's efficiency.

Make Spending Decisions to Solve Workplace Problems	7. Calculate Costs and Discounts 8. Make Consumer Comparisons

Example: As a security system installer, you may be required to calculate costs and discounts for the customers you serve.

Copyright © The McGraw-Hill Companies, Inc.

SOLVE PROBLEMS WITH WHOLE NUMBERS AND NEGATIVE NUMBERS

Addition, subtraction, multiplication, and division of whole numbers are important skills in any career cluster, and manufacturing is no exception. A cargo or freight agent, for example, must be able to add to find the total number of packages being shipped to one place.

1. You work as a production associate in a furniture plant. You have to cut a 96-inch long board into 8-inch long pieces to build a prototype of a new desk. How many pieces do you cut?

    A. 12

    B. 36

    C. 48

    D. 56

    E. 84

2. As a machine technician working at a TV manufacturing plant, one of your duties is to replace parts in faulty machines. There are 17 faulty machines and you replace parts in 9 of the machines. How many machines still need new parts?

    A. 1

    B. 7

    C. 8

    D. 17

    E. 26

3. You are a machine operator at a bottling plant. Your job is to feed a machine with empty plastic bottles to be filled with juice. It takes 30 minutes to fill a full tray of 1,000 bottles. How many minutes does it take to fill 4 trays of bottles?

    A. 30

    B. 60

    C. 90

    D. 120

    E. 150

4. You are an operator working on a DVD manufacturing line. You usually work an 8-hour shift, but today the line was down for an hour. In order to manufacture the required number of DVDs per day, your supervisor asked you to work two extra hours. How many hours were you at work today?

    A. 6

    B. 8

    C. 10

    D. 12

    E. 14

Copyright © The McGraw-Hill Companies, Inc.

5. You are a material handler in a small, local plant that makes and applies labels to bottles. The adhesive used to apply the labels must be stored at –7° C. If one of the storage room's temperature is currently 4° C, how much cooler must the room be if it is to effectively store the adhesive.

   A. –3° C

   B. 3° C

   C. 11° C

   D. 14° C

   E. 15° C

6. You are an assistant working in a small printing house. Your supervisor is worried about exceeding the monthly budget. Last month the printing house you worked at had a total of $50,192 worth of expenses and brought in a total of $42,000 revenue. What is the printing house's net gain last month?

   A. –$92,192

   B. –$42,000

   C. –$8,192

   D. $8,192

   E. $92,192

7. You work as an assistant in a garment manufacturing plant. You need to order fabric for a new garment your plant is producing. You have been given access to a bank account containing $21,000. One company charges $22,530 for the fabric you need. If you purchase the fabric from this company using the bank account, what will the balance of the bank account be afterwards?

   A. –$43,530

   B. –$22,530

   C. –$1,530

   D. $1,530

   E. $43,530

8. You work as a maintenance repairer at a large manufacturing plant. Chemical Compound A, a chemical used at the plant you work at, needs to be –5° C before it can be transported and added as an ingredient to a formula. If the temperature is currently 2° C, how much cooler does the chemical have to be for transportation and addition to the formula?

   A. –3° C

   B. 5° C

   C. 7° C

   D. 8° C

   E. 11° C

ANSWER KEY

Item 1: **A** 96 ÷ 8 =12

Item 2: **C** 17 – 9 = 8
Eight machines still need new parts

Item 3: **D** 30 × 4 = 120 minutes

Item 4: **C** 8 + 2 = 10 hours

Item 5: **C** 4 – (–7) = 11°C

Item 6: **C** 42,000 – 50,192 = –$8,192

Item 7: **C** 21,000 – 22,530 = –$1,530

Item 8: **C** 2 – (–5) =7° C

Copyright © The McGraw-Hill Companies, Inc.

USE FRACTIONS, DECIMALS, AND PERCENTAGES TO SOLVE WORKPLACE PROBLEMS

In the manufacturing industry, workers come across quantities represented in many different ways. A shipping clerk may need to calculate percentages to add shipping charges to a bill. The ability to perform workplace calculation using different forms of numbers is an important workplace skill.

1. You work as an intern at a plant that manufactures electronic devices, such as diodes. Your job is to mark down the values of the materials that make up the alloy on an electronic chart that allows only decimal numbers. If the alloy is made of 75 percent gallium, what decimal value do you write down for gallium?

   A. 0.25

   B. 25

   C. 0.75

   D. 75

   E. 1

2. You are a supervisor on a auto manufacturing assembly line. Your boss tells you that you need to increase your production numbers by a third. What does that mean in percentage?

   A. $\frac{1}{3}$%

   B. 10%

   C. 20%

   D. 30%

   E. 33%

3. You work as a quality technician assistant in a printing house. Your job is to inspect all the books printed in a day before they get shipped to the customers. By noon you have inspected $\frac{13}{20}$ of the books, and by 3 p.m. you have inspected another $\frac{1}{10}$ of the books. What is the total fraction of books inspected by 3 p.m.?

   A. $\frac{1}{10}$

   B. $\frac{1}{4}$

   C. $\frac{11}{20}$

   D. $\frac{3}{4}$

   E. $\frac{17}{20}$

4. You are a machinist working for a tire manufacturing company. A machine has broken down and requires three replacement parts. The first part costs $157.25, the second part costs $262.75, and the third part costs $100.50. What is the total cost of the three parts?

   A. $262.50

   B. $383.25

   C. $429.00

   D. $467.75

   E. $520.50

Copyright © The McGraw-Hill Companies, Inc.

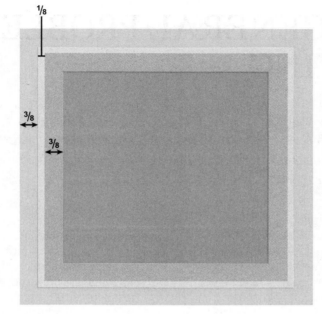

¹/₈

³/₈

³/₈

5. You work as a framing assistant in a picture framing shop. Your client brings in a square photograph. She wants to use three overlapping mats to frame the photograph, so that $\frac{3}{8}$ inch is visible from the first mat, $\frac{1}{8}$ inch is visible from the second mat and $\frac{3}{8}$ inch is visible from the third mat. What is the total visible width of each side of the mats?

A. $\frac{1}{8}$ inch

B. $\frac{3}{8}$ inch

C. $\frac{4}{8}$ inch

D. $\frac{6}{8}$ inch

E. $\frac{7}{8}$ inch

6. You work as an assistant to a mask designer who specializes in dust masks used in factories. Part of your job is to create invoices for billing customers. If the mask designer charges $40.25 per hour and he worked $7\frac{1}{2}$ hours on a mask, how much money do you bill the client?

A. $281.75

B. $286.58

C. $291.81

D. $301.05

E. $301.88

ANSWER KEY

Item 1: **C** 75% = 0.75

Item 2: **E** $\frac{1}{3}$ = 33%

Item 3: **D** $\frac{13}{20} + \frac{1}{10} = \frac{3}{4}$

Item 4: **E** 157.25 + 262.75 + 100.50 = $520.50

Item 5: **E** $\frac{3}{8} + \frac{1}{8} + \frac{3}{8} = \frac{7}{8}$ inch

Item 6: **E** 40.25 × $7\frac{1}{2}$ = $301.88

Copyright © The McGraw-Hill Companies, Inc.

USE GENERAL PROBLEM SOLVING

Some mathematical calculations require more than one operation. A quality assurance technician may need to use multiple calculations to determine the percentage of defective parts per shift. Being able to quickly perform such calculation can improve a manufacturing worker's efficiency.

1. You are a freight, stock, and material mover for a company that manufactures parts for hybrid automobile batteries. Your regular rate $25 per hour. If you work Sundays, you get paid double. How much do you get paid for working 7.5 hours on a Sunday?

   A. $50.00

   B. $187.50

   C. $202.50

   D. $375.00

   E. $750.00

2. As a quality assurance technician, you test for bacterial contamination at 200 stations in a food processing plant. During the morning, you test 75 stations. During the afternoon, you test another 80 stations. How many stations do you have left to test?

   A. 45

   B. 50

   C. 65

   D. 75

   E. 80

3. You are a machinist at a tool and die factory. The project you're working on requires several stages in order to be completed. You agree that the client can pay a deposit of $2,200 and the rest in 5 equal installments of $1,000, each upon the completion of a stage. How much does the client pay for the entire project?

   A. $1,000

   B. $2,200

   C. $5,000

   D. $6,000

   E. $7,200

4. You work as a quality technician assistant in an auto plant. As part of your job, you take random cars for test drives to see how their engines behave at various speeds. Each day you have to test 13 cars. If you work 5 days a week, how many cars do you test over 9 weeks?

   A. 27

   B. 45

   C. 65

   D. 585

   E. 615

Copyright © The McGraw-Hill Companies, Inc.

5. As a contract industrial maintenance technician, you repair industrial equipment for various manufacturing clients. You just completed a project for a client, for which you charged $600. The client calls you back with additional requests for the same project. You charge $30 per additional hour. If you work 5 additional hours, how much do you get paid, in total, for that project?

   A. $150

   B. $300

   C. $450

   D. $600

   E. $750

6. You work in the packaging department of a small bookstore. Your job is to verify the contents of the order before it is sent to shipping. The invoice for your current order shows that a customer ordered 20 copies of a book that costs $18 each and the shipping and handling is $7. You have verified the number of books in the shipment, and now you need to verify the invoice. How much should the client be invoiced?

   A. $177

   B. $218

   C. $254

   D. $367

   E. $478

7. As a production manager of a small facility manufacturing circuit boards, you need to approve the purchase of any new tools. Your electrician comes in with his research on the best prices for a replacement set of tools. His research shows that the tools he needs cost $1,120. The electrical supply store offers a 25% discount from the original price, and in addition, you get a $300 mail-in rebate. What is the final price of the tools?

   A. $460

   B. $540

   C. $625

   D. $785

   E. $830

8. As a material handler on a ball-point pen manufacturing line, you have to monitor how many pens are to be produced for various projects. You have to have 50,000 pens by tomorrow. In the morning 25,000 pens were produced and in the afternoon 12,570 pens were produced. How many more pens need to be produced?

   A. 11,570 pens

   B. 12,000 pens

   C. 12,430 pens

   D. 13,970 pens

   E. 15,000 pens

ANSWER KEY

Item 1: **D** $25 \times 2 = 50$; $50 \times 7.5 = \$375.00$

Item 2: **A** $200 - 75 - 80 = 45$ stations

Item 3: **E** $5 \times 1,000 = 5,000$; $5,000 + 2,200 = \$7,200$

Item 4: **D** $9 \times 5 = 45$; $45 \times 13 = 585$ cars

Item 5: **E** $5 \times 30 = 150$; $600 + 150 = \$750$

Item 6: **D** $20 \times 18 = 360$; $360 + 7 = \$367$

Item 7: **B** $1,120 - 1,120 \times 25\% = 840$; $840 - 300 = \$540$

Item 8: **C** $25,000 + 12,570 = 37,570$; $50,000 - 37,570 = 12,430$ pens

Copyright © The McGraw-Hill Companies, Inc.

SOLVE PROBLEMS IN GEOMETRY

Knowing how to determine the perimeters and areas of objects and spaces is an important skill in the manufacturing industry. A tool and die maker may need to determine the perimeters and areas of an object to make parts that are the proper size. Parts may be in many shapes, including circles and rectangles.

1. You are a factory manager working at a factory that manufactures speakers. You wish to install a public address system in the factory. To do this, you need to run wire around the perimeter of a large room. The room is 113 feet long and 57 feet wide. What is the perimeter of the room?

   A. 114 feet

   B. 160 feet

   C. 170 feet

   D. 340 feet

   E. 6,441 feet

2. As a carpenter working in a furniture factory, you design a round dining room table with a diameter of 4 feet. In order to figure out how much material is needed to construct the table, you need to compute the area of the table. What is the area of the dining room table?

   A. 12.56 square feet

   B. 14.00 square feet

   C. 23.74 square feet

   D. 36.00 square feet

   E. 45.96 square feet

3. You are a quality inspector working at an automobile piston manufacturing plant. If the diameter of the part is 5 inches, what is the circumference?

   A. 11.20 inches

   B. 12.30 inches

   C. 13.40 inches

   D. 14.50 inches

   E. 15.70 inches

4. You are a robotics technician working for a company that produces toys. The company you work for wants to purchase new robotic machines to help aide toy production. One machine you are considering purchasing has a rectangular base that is 10 feet long and 4.5 feet wide. What is the area of the base of the machine?

   A. 14.5 square feet

   B. 29 square feet

   C. 45 square feet

   D. 50 square feet

   E. 60 square feet

Copyright © The McGraw-Hill Companies, Inc.

5. As a designer at a molded swimming pool manufacturing plant, you work on the landscape of your clients' properties. The client wants a rectangular 12 × 20-foot pool on the property. You need to determine the area of the pool to determine how much material will be needed for it's production. What is the area of the pool?

A. 32 square feet

B. 64 square feet

C. 70 square feet

D. 240 square feet

E. 480 square feet

6. You work as a framing assistant in a framing factory. A client comes in with an order for frames for prints of a famous painting. The prints are 22 inches long and 17 inches wide. What is the area of the print?

A. 39 square inches

B. 78 square inches

C. 100 square inches

D. 160 square inches

E. 374 square inches

7. You work as a measuring specialist at a carpet manufacturer that sells directly to customers. You measure a room for a client who will be installing the carpet. One corner of the room will not have any carpeting. Based on the diagram, how many square feet of carpeting will be needed to carpet the room?

A. 100 square feet

B. 110 square feet

C. 120 square feet

D. 140 square feet

E. 160 square feet

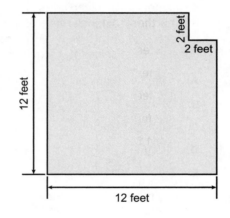

ANSWER KEY

Item 1: **D** 113 + 57 = 170; 170 × 2 = 340 feet

Item 2: **A** 4 ÷ 2 = 2; 3.14 × 2² = 12.56 square feet

Item 3: **E** 3.14 × 5 = 15.70 inches

Item 4: **C** 10 × 4.5 = 45 square feet

Item 5: **D** 20 × 12 = 240 square feet

Item 6: **E** 22 × 17 = 374 square inches

Item 7: **D** 12² = 144; 2² = 4; 144 − 4 = 140 square feet

Copyright © The McGraw-Hill Companies, Inc.

CALCULATE WITH CONVERSIONS AND FORMULAS

Some calculations in manufacturing may require using conversions and formulas. As a solderer, you may need to use formulas to determine the time it takes to complete a project. You may also need to convert units when measuring parts and materials.

1. You are a line operator soldering connectors on a side station in an auto parts manufacturing plant. You spend 5 minutes on each connector—placing the parts in the machine, soldering them, and inspecting for errors. How much time does it take to solder 15 connectors?

    A. 1 hour 15 minutes

    B. 2 hours

    C. 3 hours 30 minutes

    D. 4 hours

    E. 5 hours 20 minutes

2. You are a machine inspector. You need to test a recently programmed machine that counts and sorts coins to make sure it works in a real-world scenario. You test it by inserting $3.59 worth of change. Which of the following combinations of coins would be the correct display?

    A. 11 quarters, 6 nickels, and 9 pennies

    B. 10 quarters, 8 dimes, and 2 pennies

    C. 12 quarters, 6 dimes, and 4 pennies

    D. 9 quarters, 4 dimes, and 6 pennies

    E. 12 quarters, 5 dimes, and 9 pennies

3. You are a factory manager working for a kitchen cabinet plant. During the morning you have a meeting that lasts 30 minutes. In the afternoon, you spend 3 hours and 40 minutes in meetings. How much time do you spend in meetings throughout the day?

    A. 30 minutes

    B. 40 minutes

    C. 3 hours

    D. 3 hours 40 minutes

    E. 4 hours 10 minutes

4. You work as a line operator in a bottling plant. Your job is to sort bottles by size. You sort three 500-milliliter bottles, two 750-milliliter bottles and one 1-liter bottle. Each kind of bottle will be filled with a carbonated drink. There are 1000 milliliters in a liter. How many liters of carbonated drink will fit in the bottles you just sorted?

    A. 1 Liter

    B. 1.5 Liters

    C. 2.5 Liters

    D. 4 Liters

    E. 5.5 Liters

Copyright © The McGraw-Hill Companies, Inc.

5. As a framing assistant, you help customers envision how their artwork would looked framed. A customer wants to have 2 frames done: one with a total frame material length of 12 feet, 9 inches and another with a total of 14 feet, 10 inches. You have 28 feet of the frame stock on hand. How much frame material do you have left over after both frames have been made?

   A. 4 inches

   B. 5 inches

   C. 6 inches

   D. 7 inches

   E. 8 inches

6. As a communications technician, you need to travel to the site of a new cell phone tower to test the custom parts your company manufactured for the tower. You have to drive 156 miles, and you want to be there on time, at 9 a.m. You drive at an average speed of 65 miles per hour. How long does it take you to get to your client?

   A. 1 hour

   B. 1.5 hours

   C. 2 hours

   D. 2.4 hours

   E. 3 hours

7. You are a quality assurance technician working at an auto brake manufacturing plant. You have to check 2 percent of the 2,000 brake valves produced per shift. There are three 8-hour shifts and each produces between 1,700 and 2,300 brake valves, depending on downtime. How many valves do you check during your shift?

   A. 17

   B. 20

   C. 23

   D. 30

   E. 40

8. You are a supervisor in an LED light bulb manufacturing plant. You work two 10-hour shifts. During the first shift, your line produces 4,000 lightbulbs. During the second shift, it produces 25 percent more. During the two shifts, the line produces 1,850 lightbulbs more than in the previous day. How many lightbulbs did the line produce during both the first and second shift?

   A. 7,000

   B. 8,000

   C. 9,000

   D. 10,000

   E. 11,000

ANSWER KEY

Item 1: **A** 5 × 15 = 75; 75 minutes = 1 hour 15 minutes

Item 2: **E** 4 quarters × 3 = 12 quarters in 3 dollars; 5 dimes × 10 cents = 50 cents; 9 pennies × 1 cent = 9 cents. The correct answer is 12 quarters, 5 dimes, and 9 pennies.

Item 3: **E** 30 minutes + 3 hours 40 minutes = 4 hours 10 minutes

Item 4: **D** 500 ÷ 1,000 = 0.5; 0.5 × 3 = 1.5; 750 ÷ 1,000 = 0.75; 0.75 × 2 = 1.5; 1.5 + 1.5 + 1 = 4 liters

Item 5: **B** 12 feet 9 inches + 14 feet 10 inches = 27 feet 7 inches; 28 feet − 27 feet 7 inches = 5 inches

Item 6: **D** 156 ÷ 65 = 2.4 hours

Item 7: **E** 2,000 × 2% = 40 valves

Item 8: **C** 4,000 + (4,000 × 25%) = 5,000; 4,000 + 5,000 = 9,000 lightbulbs

Copyright © The McGraw-Hill Companies, Inc.

MANIPULATE FORMULAS TO SOLVE PROBLEMS

For some calculations in manufacturing, a formula may need to be manipulated to solve a problem. For example, a robotics engineer may need to manipulate a formula to calculate the distance a robotic arm will move an object. Workers should be able to work with formulas to find the information required.

1. As a computer-aided design technician for a fuel cell manufacturing company, you are designing a rectangular box with a volume of 6,000 cubic centimeters for your company's new smaller fuel cell design. The base of the box has a width of 20 centimeters and a length of 30 centimeters. What is the height of the box?

   A. 10 centimeters

   B. 15 centimeters

   C. 20 centimeters

   D. 25 centimeters

   E. 30 centimeters

2. You are a plastic mold operator at a company manufacturing hard plastic smartphone cases. The hard plastic case has a rectangular shape. You have to affix labels to the back of the case. How long are the labels, if they are 2.5 inches wide and have an area of 11.25 square inches?

   A. 4.5 inches

   B. 5.0 inches

   C. 5.5 inches

   D. 6.0 inches

   E. 6.5 inches

3. You are a lab technician testing car engines. You test a car to find out how long it takes to reach a target speed, starting from a standstill. The average power it takes to reach the target speed is 15KWatts and the kinetic energy needed to reach the target speed is 225KJoules. Using the formula for average power (average power = kinetic energy/time in seconds), how long does it take the car to reach the targeted speed?

   A. 1 seconds

   B. 9 seconds

   C. 15 seconds

   D. 22.5 seconds

   E. 135 seconds

4. You are a factory manager working for a company that produces microchips. You are having new low VOC floors installed in one of the work rooms to reduce dust and fumes in the room. The width of the room is 30 feet and the perimeter of the room is 140 feet. What is the length of the room?

   A. 30 feet

   B. 40 feet

   C. 50 feet

   D. 60 feet

   E. 70 feet

Copyright © The McGraw-Hill Companies, Inc.

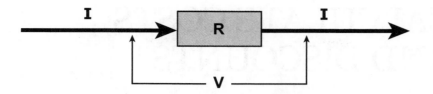

Ohm's Law: Voltage (Volts) = Current (amps) x Resistance (ohms)

5. You are an electrical technician testing an electric board you built for a project simulation. You have to replace a resistor, but first you re-measure the current that goes through it and the voltage, to make sure you replace it with the right value of resistance. The current reads 6 amps and voltage reads 12 volts. Note that voltage is equal to the current times the resistance. What should be the value of the new resistance?

A. 2 Ohms

B. 3 Ohms

C. 4 Ohms

D. 5 Ohms

E. 6 Ohms

6. As an electrical quality assurance technician working for a wind turbine manufacturing company, you test a new wind energy installation. You measure 96-watts of power on the installation's panel. What voltage do you measure when it is delivers an electric current of 8 amps?
Formula of the electric power (measured in watts) is: Power (watts) = Voltage (Volts) × Current (Amps) or P = V × I

A. 8 volts

B. 12 volts

C. 16 volts

D. 20 volts

E. 24 volts

ANSWER KEY

Item 1: **A** 6,000 cubic centimeters = 30 centimeters × 20 centimeters × height = 600 square centimeters × height; divide to find the height: 6,000 cubic centimeters ÷ 600 square centimeters = 10 centimeters = height

Item 2: **A** (11.25 square inches) ÷ 2.5 in = 4.5 inches

Item 3: **C** 225KJ ÷ 15KW = 15 seconds
It takes the car 15 seconds to reach the target speed, also called velocity.

Item 4: **B** 140 ÷ 2 = 70; 70 − 30 = 40 feet

Item 5: **A** 12 ÷ 6 = 2 Ohms

Item 6: **B** 96 ÷ 8 = 12 volts

Copyright © The McGraw-Hill Companies, Inc.

7

CALCULATE COSTS AND DISCOUNTS

Perform Basic
Arithmetic
Calculations
to Solve
Workplace
Problems

Apply
Computations
to Solve
Workplace
Problems

Solve
Measurement
Problems

**Make
Spending
Decisions
to Solve
Workplace
Problems**

Some jobs in the manufacturing industry require workers to calculate costs and discounts. A plumber, for example, needs to be able to calculate a mark-up that will bring in a profit while being acceptable to the client.

1. As a manager at a paper recycling facility, you buy recycled cardboard boxes. Your supplier offers 1,000 boxes (minimum purchase) at $0.50 a box plus a 7% sales tax. How much do you end up paying for the recycled cardboard boxes?

    A. $515

    B. $520

    C. $525

    D. $530

    E. $535

2. You are a quality assurance assistant working in a semiconductor manufacturing plant. The supervisor just announced the introduction of a new line, using an innovative method of manufacturing semiconductors. To learn more about the new technology about to be introduced, you order a book from an online vendor. The book costs $30. Shipping is 3% of your purchase, but sales tax is free.How much do you end up paying for the book?

    A. $29.25

    B. $29.50

    C. $30.25

    D. $30.90

    E. $32.75

3. You are the manager at a small bottling company specializing in craft beers. You are in need of IT support to help maintain a customer database. You have decided to go with a company that offers support for $1,000 a month. The contractor also has a 20% discount for the first 4 months to first-time customers who sign a 1-year contract. What is the total amount you would have to pay for IT support for the first year?

    A. $1,000

    B. $1,200

    C. $3,200

    D. $8,000

    E. $11,200

4. As a manager's assistant at a paint manufacturing plant, you are responsible for ordering marketing materials for the company. Your vendor has raised the shipping and handling fees by 8% for orders under $100. The fee was $5.50. Your order is for $95 of marketing materials. What is the total cost for your order?

    A. $100.94

    B. $101.15

    C. $101.25

    D. $101.35

    E. $101.45

Copyright © The McGraw-Hill Companies, Inc.

5. You are a manager at a plastic frame manufacturing plant. After taking inventory, you decide to offer the frames that have some small errors at a discounted price. You announce a 76% sale on the rejected frames for the weekend only. If the discounted price is $5.40 for a frame, what is the cost of the frame before discount?

A. $20.00

B. $21.60

C. $22.50

D. $23.40

E. $24.50

6. You are a window cutter in a small window manufacturing business. During the holidays, you offered a store-wide discount. The discount expired in January. It costs $8 to cut a window during the holidays and this is 80% of the standard price. How much does it cost to cut a window in January?

A. $6

B. $8

C. $10

D. $12

E. $14

7. As a manager at a small distillery, you sell your small-batch vodkas in your factory store and online. You offer free shipping for first-time customers, but you charge 7% sales tax. A small shop buys $4,500 worth of your bottles. What is the total cost of the order?

A. $4,655

B. $4,680

C. $4,750

D. $4,775

E. $4,815

8. You are a manager working at a shelving manufacturing plant that uses steel to create industrial shelves for warehouses. Part of your job is to purchase steel. You can buy it from a steel recycler for $420 per ton, with a shipping cost of 20%. How much does it cost for one ton of steel?

A. $336

B. $420

C. $440

D. $504

E. $545

ANSWER KEY

Item 1: **E** $1,000 \div .50 = 500$; $500 + (500 \times 7\%) = \$535$

Item 2: **D** $30 + 30 \times 3\% = \$30.90$

Item 3: **E** $4 \times 1,000 = 4,000$; $4,000 - (4,000 \times 20\%) = 3,200$; $12 - 4 = 8$; $8 \times 1,000 = 8,000$; $3,200 + 8,000 = \$11,200$

Item 4: **A** $5.50 + 5.50 \times 8\% = 5.94$; $95 + 5.94 = \$100.94$

Item 5: **C** $100\% - 76\% = 24\%$; $5.40 \div 24\% = \$22.50$

Item 6: **C** $80\% = 0.08$; $8 \div 0.8 = \$10$

Item 7: **E** $4,500 + (4,500 \times 7\%) = \$4,815$

Item 8: **D** $420 + (420 \times 20\%) = \$504$

Copyright © The McGraw-Hill Companies, Inc.

8

Perform Basic
Arithmetic
Calculations
to Solve
Workplace
Problems

Apply
Computations
to Solve
Workplace
Problems

Solve
Measurement
Problems

**Make
Spending
Decisions
to Solve
Workplace
Problems**

MAKE CONSUMER COMPARISONS

Manufacturing workers who make purchasing decisions or recommendations must often make calculations that compare two purchasing options. A purchasing agent for a manufacturer may ask for bids to provide raw materials and select the one with the best price. Being able to make these calculations and find the best deal is an important skill in this industry.

1. As an office manager at a small precast concrete manufacturing company, you order business cards for your boss. Company A offers boxes of 100 business cards for $25. Company B offers boxes of 250 business cards for $50. Which company offers the best price per business card?

   A. Company B, $0.20 per business card

   B. Company A, $0.25 per business card

   C. Company A, $0.20 per business card

   D. Company B, $0.25 per business card

   E. Company A, $25 per box

2. You are a logistician at a printer manufacturing company. Before any printer can be shipped, it has to be tested several times. You need to purchase 50,000 pieces of paper for these tests. Supplier A sells a box of 1,000 pieces of paper for $20 and charges $13 for shipping. Supplier B sells a box of 2,500 pieces of paper for $51 and provides free shipping. What is the least amount you can spend on the 50,000 pieces of paper?

   A. $987

   B. $1,013

   C. $1,017

   D. $1,020

   E. $1,033

3. You are the manager of a small tool and die company. You want to enroll the business in a new cell phone plan that also offers text messaging. Your first bid is $65 per month for phone and data, plus 200 text messages for an additional $10. How much does each individual text message cost with this plan?

   A. $0.02

   B. $0.05

   C. $0.33

   D. $0.38

   E. $0.44

4. You are an electrician working in a furniture manufacturing plant. You need to purchase electrical wire for your current project. You find a 500-foot roll of wire for $37.95, a 100-foot roll for $11.95, and a 50-foot roll for $7.95. Which roll should you purchase if you want the cheapest rate per foot and what is that rate?

   A. 500-foot roll, $0.08 per foot

   B. 100-foot roll, $0.12 per foot

   C. 100-foot roll, $0.16 per foot

   D. 100-foot roll, $0.55 per foot

   E. 50-foot roll, $0.16 per foot

Copyright © The McGraw-Hill Companies, Inc.

5. You work as the plant manager for an organic cotton cloth manufacturing plant. Your supervisor asks you to research the prices of a new weaving loom. The price for the size and style of the loom you need is listed at $1,500 by three different vendors, but each vendor is offering a discount. Vendor 1 offers 20 percent off the original price and free shipping and handling. Vendor 2 offers a discount of a third off the original price but is charging $127 for shipping and handling. Vendor 3 offers 10 percent discount from the original price, free shipping and handling, and a $300 mail-in rebate. What is the best price that you can get for the loom?

A. Vendor 3 at $1,050

B. Vendor 2 at $1,127

C. Vendor 3 at $1,185

D. Vendor 1 at $1,200

E. Vendor 2 at $1,500

6. You are an office manager for a blinds and drapery manufacturing facility. Part of your job is to oversee the renovations of the office space. You need to install new carpet in a rectangular room that is 18 feet by 20 feet, and you get three bids. Bid 1 is $20 per square yard plus a $125 installation fee. Bid 2 is $2.75 per square foot installed. Bid 3 is a flat rate of $1,000 and you receive a 10% discount. What is the best offer to replace the office carpet?

A. Bid 1 at $800

B. Bid 3 at $900

C. Bid 1 at $925

D. Bid 2 at $990

E. Bid 3 at $1000

ANSWER KEY

Item 1: **A** Company A: 25 ÷ 100 = 0.25; Company B: 50 ÷ 250 = 0.20; Company B, at $0.20 per business card, has the better price

Item 2: **B** Supplier A: 50,000 ÷ 1,000 = 50; 50 × 20 = 1,000; 1,000 + 13 = 1,013; Supplier B: 50,000 ÷ 2,500 = 20; 20 × 51 = 1,020; $1,013 is the least amount your can spend for 50,000 pieces of paper.

Item 3: **B** 10 ÷ 200 = $0.05

Item 4: **A** First roll: 37.50 ÷ 500 = 0.08; Second roll: 11.95 ÷ 100 = 0.12; Third roll: 7.95 ÷ 50 = 0.16; The first roll is the cheapest at $.08 per foot.

Item 5: **A** Vendor 1: 1,500 − 1,500 × 20% = 1,200; Vendor 2: 1,500 − 1,500 × $\frac{1}{3}$ = 1,000; 1,000 + 127 = 1,127; Vendor 3: 1,500 − 1,500 × 10% = 1,350; 1,350 − 300 = $1,050; at $1,050, vendor 3 has the best price.

Item 6: **B** Bid 1: 20 × 18 = 360; 360 ÷ 9 = 40; 40 × 20 = 800; 800 + 125 = 925; Bid 2: 360 × 2.75 = 990; Bid 3: 1,000 − (1,000 × 10%) = $900; bid 3 at $900 is the best offer.

Copyright © The McGraw-Hill Companies, Inc.

CAREER CLUSTERS AND PATHWAYS

A **career cluster** is a grouping of jobs and industries based on common characteristics. A **career pathway** is an area of focus within a career cluster. You can explore each of the following career clusters and pathways in McGraw-Hill Workforce's *Career Companion* series.

Agriculture, Food & Natural Resources
Food Products and Processing Systems
Plant Systems
Animal Systems
Power, Structural & Technical Systems
Natural Resources Systems
Environmental Service Systems
Agribusiness Systems

Architecture & Construction
Design/Pre-Construction
Construction
Maintenance/Operations

Arts, Audio/Video Technology & Communications
Audio and Video Technology and Film
Printing Technology
Visual Arts
Performing Arts
Journalism and Broadcasting
Telecommunications

Business Management & Administration
General Management
Business Information Management
Human Resources Management
Operations Management
Administrative Support

Education & Training
Administration and Administrative Support
Professional Support Services
Teaching/Training

Finance
Securities & Investments
Business Finance
Accounting
Insurance
Banking Services

Government & Public Administration
Governance
National Security
Foreign Service
Planning
Revenue and Taxation
Regulation
Public Management and Administration

Health Science
Therapeutic Services
Diagnostic Services
Health Informatics
Support Services
Biotechnology Research and Development

Hospitality & Tourism
Restaurants and Food/Beverage Services
Lodging
Travel & Tourism
Recreation, Amusements & Attractions

Human Services
Early Childhood Development & Services
Counseling & Mental Health Services
Family & Community Services
Personal Care Services
Consumer Services

Information Technology
Network Systems
Information Support and Services
Web and Digital Communications
Programming and Software Development

Law, Public Safety, Corrections & Security
Correction Services
Emergency and Fire Management Services
Security & Protective Services
Law Enforcement Services
Legal Services

Manufacturing
Production
Manufacturing Production Process Development
Maintenance, Installation & Repair
Quality Assurance
Logistics & Inventory Control
Health, Safety and Environmental Assurance

Marketing
Marketing Management
Professional Sales
Merchandising
Marketing Communications
Marketing Research

Science, Technology, Engineering & Mathematics
Engineering and Technology
Science and Math

Transportation, Distribution & Logistics
Transportation Operations
Logistics Planning and Management Services
Warehousing and Distribution Center Operations
Facility and Mobile Equipment Maintenance
Transportation Systems/Infrastructure Planning, Management and Regulation
Health, Safety and Environmental Management
Sales and Service

Copyright © The McGraw-Hill Companies, Inc